Praise for *The Pearl and the Flame* . . .

"This book distills a lifetime of serious thinking about both Judaism and ecology. Margalit reads and dialogues equally well with Wendell Berry and the Rabbi of Piasecne. A fresh and inspiring perspective for the special times in which we live."

— Rabbi Arthur Green, Rector, Hebrew College Rabbinical School, author of *Judaism for the World*

"Natan Margalit's innovative book, *The Pearl and the Flame*, explores how ancient Jewish practices embody organic ways of thinking and doing. Now that technology and reductionist thinking have brought us to the edge of extinction, what we humans most need is to find ways to relate to the whole. . . . In a society where community and meaning are fraying, Judaism's ancient ways can offer not only connection but a path to sustainability. Margalit's book, full of down-to-earth personal stories as well as astute cultural observations, beautifully strings the pearls of Judaism and ecological thinking together to create a relevant and nourishing whole."

— Rabbi Jill Hammer, Ph.D,. author, *The Jewish Book of Days: A Companion for All Seasons*

"There aren't many rabbis who can weave together pieces of wisdom from Wendell Berry, Mary Douglas, and Kalonymous Kalman Shapira. It reflects the genius of a book that is an antidote to an ever more fragmented world. Reb Natan's vision of integration, of cosmic wholeness, of, what he calls "the Holy Source of All" is the kind of Torah we need in our world today."

— Rabbi Sid Schwarz, Senior Fellow, Hazon; author of *Judaism and Justice: The Jewish Passion to Repair the World*

"Margalit has written a rich, engaging, and powerful book. As I read *The Pearl and the Flame*, I felt that I was sitting at the feet of a sage. It is filled with gems, among which were, for me, the focus on patterns/systems, Jewish teachings about relationships, cultures of control vs. cultures of relationship, and more. Much of what he writes resonates with Buddhist teachings and my own interests."

— Christopher Ives, Ph.D., Professor of Religious Studies, Stonehill College, author of *Zen on the Trail: Hiking as Pilgrimage*

"I am grateful to Rabbi Natan Margalit for writing a book that is both comprehensive and profound, connecting me through the wide reaches of sources from ancient to modern, to a path that is illuminated by the flame of truth and made beautiful by pearls of wisdom. *The Pearl and The Flame* is a treasure of erudite scholarship and personal insight that will delight as much as it informs."

— Rabbi Shefa Gold, author, *Torah Journeys: The Inner Path to the Promised Land*

"Margalit's deeply personal, *The Pearl and the Flame*, identifies holes in the modern soul and world that need to be addressed and, as remedies, offers up the wisdom of the ages found in Jewish writings and traditions, stressing connectedness, relationship, care, and emergence. It speaks to the same core of ideas that my own work has involved, but with a close tie to Jewish wisdom."

— John Ehrenfeld, former Director of the Program in Technology, Business and the Environment at M.I.T., author, *The Right Way to Flourish: Reconnecting with the Real World*

"Rabbi Margalit reveals the root cause of all our suffering, environmental and social: the mistaken notion that we humans are *apart from* rather than *a part of* a greater wholeness. Drawing on contemporary science and ancient wisdom, *The Pearl and the Flame*, introduces us to a vibrant Judaism that overcomes this existential error and offers us a way to embrace all life in the greater Aliveness that is God."

— Rabbi Rami Shapiro, author of *Judaism Without Tribalism*

"Gorgeous and amazing! The simple act of reading this book— so filled with life-changing insights and new takes on an old tradition—just may change your life."

— Rabbi Benay Lappe, Founder and Rosh Yeshiva of SVARA: A Traditionally Radical Yeshiva

The *Pearl*
and the *Flame*

A Journey into Jewish Wisdom
and Ecological Thinking

Natan Margalit

Albion
Andalus
Boulder, Colorado
2022

*"The old shall be renewed,
and the new shall be made holy."*
— Rabbi Avraham Yitzhak Kook

Albion-Andalus, Inc.
P. O. Box 19852
Boulder, CO 80308
www.albionandalus.com

Design and composition by Albion-Andalus Books
Cover design by D.A.M. Cool Graphics
Author photo by Thea Breite

ISBN: 978-1-953220-18-9 (Hardcover)
ISBN: 978-1-953220-39-4 (Paperback)

Manufactured in the United States of America

To my mother, who brought me into this world,
and to my wife, Ilana, who has made a world with me,
and to our children, Nadav and Eiden,
who are the world that is coming.

Contents

PART ONE:

Our Relation to Creation

PART TWO:

Minyan—Emergence and Life

PART THREE:

Mikdash—The Concentric Circles of Life

PART FOUR:

Mitzvah—The Spiritual Technology of Change

Acknowledgements

This book is the fruit of a tree with many branches and with roots drawing nourishment from many sources. I know I will be missing some people, either through forgetfulness or lack of space, who have made important contributions to its ripening. I offer my apologies and my thanks, in any case. I will also thank people who surely will not agree with everything I say or who may feel I've misunderstood their teachings or contributions. I take responsibility for the opinions I express in the book as my own interpretations and any mistakes as my own, and I only wish to offer my deep gratitude for all who have helped me along the way.

I want to thank my teachers at the Pardes Institute of Jewish Studies from the early years in the '80s, especially Dov Berkovits, who remains a major source of inspiration, and Avraham Walfish, whose academic work in the Mishnah has helped shift the discussion and paved the road for a deeper understanding of rabbinic texts. All the Pardes teachers and my fellow students in those magical days and hours in the beit midrash in Baka'a shaped my outlook on life in so many ways.

Many teachers have been influential to my growth including Avivah Zornberg, Rav Sha'g'ar (Shimon Gershon Rosenberg, of blessed memory), Yehuda Gellman, Arthur Green, Daniel Boyarin, Shefa Gold, and Sylvia Boorstein, to name a few.

I want to thank some of the teachers, friends and colleagues who really saw me even when I wasn't making myself easily seen, who believed in me even if I didn't fit the mold, starting from my kindergarten teacher, Mrs. Kirkpatrick, and my seventh-grade social studies teacher (mentioned in the book) Mr. Barry Yamamoto, including my brilliant and tough as nails anthropology

The *Pearl* and the *Flame*

professor and thesis advisor who nevertheless saw me and supported me: Professor Gail Kelly, of blessed memory, Marcia Praeger, Dean (now emeritus) of the Aleph Ordination Program, SooJi Min-Maranda, Executive Director of ALEPH, and of course, Reb Zalman Schachter-Shalomi, of blessed memory.

I want to thank a revered teacher, not of this generation, who simultaneously has made me feel that it would be a miracle if I were to someday reach even to his ankles in spiritual service, courage and insight, and still has lifted me, inspired me and given me so much joy in learning his holy Torah: Rabbi Kalonymus Kalman Shapiro, the Piaseczner Rebbe (may the memory of the righteous and the martyrs be a shield and a blessing for us).

I'd like to thank some of the *melakhim* (messengers of the Divine) who have shown up in my life and connected me to what I needed, be it a house, a car, my wife (!), medicine, chocolate or a teacher: Azriel Cohen, of blessed memory, Marc Grossman, and Daniel Orlansky.

I want to thank all the people who helped, over many years, to read drafts and listen to ideas, help craft proposals to publishers, and make introductions to agents or publishers: especially to my old friend from college, Samuel Fromartz who worked with me in the early drafts and who made introductions to his agent and publisher. To Tirzah Firestone who (re)introduced me to Netanel Miles-Yépez, who agreed to publish this book with Albion-Andalus. To Jeremy Sher who helped me start the organization Organic Torah which was the vehicle for the many blog posts which helped me explore the ideas in this book and also to develop my non-academic writing voice. To Benyamin Lichtenstein, Djana Paper, Marilyn Paul, Lisa Tener, David Jaffe, Mark McDiarmid, Jill Hammer, and Christopher Ives who read drafts and offered helpful comments and corrections. To friends, family and colleagues who have generously given advice, moral, spiritual networking and financial support: Bracha Leah Zeitler, Janee Graver, Joel Segel, Daniel Berman, Amos Lassen, Ebn Leader, Aubrey Glazer, Jordan Wolfson, my in-laws Beth and

Julian (Juggy to us) Gott. My friends and colleagues at the Aleph Ordination Program and at Aleph for their encouragement and support. To Jamelah Zidan, who did a great job on final editing and preparing the book for publication and to Netanel Miles-Yépez for working with me in bringing this book to light with Albion-Andalus Books. And to all the others who I am inadvertently missing or who there isn't space to name, I appreciate the support, guidance and encouragement of so many along the way.

I want to thank all my students who have sharpened my thoughts with your questions, encouraged me with your appreciation and given me so many insights in our discussions. I could never have come up with all this on my own: this book is truly an example of emergence.

I want to thank some of the path finders in Jewish environmental movement: Arthur Waskow, Fred Scherlinder Dobb, Ellen Bernstein, David Seidenberg, Michal Smart, Michael Comins, Nigel Savage, Adam Berman, Lawrence Troster, of blessed memory, Jeremy Benstein, Nili Simchai, Shamu Sadeh, to name some with whom I've been most connected.

To my siblings, Karen, Dan and Laura: you're my home base and I love the easy comfort, humor and commitment to each other that we share. To my parents for all their patience and loving support through those many years of searching and exploring, trying and falling down and trying again. Always on my side and always giving and generous. My father, Herbert Margulies, of blessed memory, who modeled integrity, caring responsibility and good judgement, and to my mother, Fran Margulies, who taught me (and still models for me) how to care for my body with daily outdoors exercise, good food, and community, and who shares my interests in Judaism, in finding dynamic balances, and in good writing: My supporter and believer from the very beginning.

And, finally, to my wife, Ilana: the fact that we found each other is still the best proof I have that there is a God who loves us all. Marrying you was the best decision I've ever made. Your patience, loving support, deep intuition and understanding have kept me on the path and true to my creative goals, balancing our equally creative and holy, but sometimes more mundane goals of raising our kids and figuring out how to make a living.

And to our children, Nadav and Eiden, who remind me that I can no longer even aspire to being cool, but instead I've become a dad, in my jokes, my clothing and my musical taste. Oh, well. I'll take it, and all that goes along with being your dad. You bring joy to me that can't be expressed in words. Keep growing into your unique selves, and don't forget the message in between the lines of this book: as it is written (Proverbs 24:16) "The righteous person falls seven times, and gets back up..." In other words, never give up. You are, if not the only, two of the biggest, reasons that I hope and work for a better future for our world.

Preface
Beyond Either/Or

Twenty years old and wandering through Israel in search of my roots, I was the perfect bait for the ultra-religious recruiters plying the streets of Jerusalem. One day, as I waited at the Jerusalem bus station, a tall, thin man dressed in the black clothing of the very religious approached me. His striking tan contrasted with the pallor usually associated with the religious men I had been seeing in certain Jerusalem neighborhoods. I wondered if he spent many hours in the sun. I later learned that he did spend much of his time walking the streets of Jerusalem, looking for young Jews like me.

His name was Rabbi Meyer Schuster, and he worked for the new *ba'al t'shuvah yeshivah*—Aish HaTorah, a school of Jewish learning specially adapted for bringing secular Jews back to "Torah True Judaism," a right-wing variety of Orthodoxy. He did his job well, with a few pick-up lines designed for different types of young seekers. For someone who looked as if they had just gotten off the plane from India, he said, "Do you want to come hear a Jewish holy man?" To me he said, "Do you want to come hear a lecture on Jewish philosophy?"

I got in a cab with him and went to the Old City through winding alleys to the yeshivah, where I heard a lecture by the head of the institution, Rabbi Noach Weinberg. Rabbi Noach was probably in his late forties, with a full, graying beard. Radiating charisma, he spoke to the class about how to find happiness and about looking for goodness in people, rather than in material things. His entertaining style and appealing message stick with me in detail to this day.

The rabbi kept returning to the evils of the Western world: its shallowness and false idols of sex, fame, and money. He presented the Torah, in contrast, as a perfect, divine gift of wisdom, revealed directly from God and carefully guarded by—you guessed it— people like him. After the class a young man greeted me. He was about my age and wore the uniform of the initiates: white shirt and black pants, jacket, and velvet yarmulke.

He guided me to another gray-bearded rabbi who invited me to join him at his table in the corner and talk. "You are at the crossroads of your life," he said. "You owe it to yourself to stay here at least a couple of weeks. Just listen. Ask questions. You can argue as much as you want, and only accept what makes sense to you."

I found myself in deep inner turmoil. I felt attracted, because this place offered what I had been looking for: meaning, belonging. Yet, something didn't feel quite right. The deck seemed stacked against me. How was I, a 20-year-old kid, supposed to win an argument with these 50-year-old rabbis who have been doing this their whole lives? They were going to run circles around me and convince me to join their cult!

Fear gripped me. It shocked me that they rejected my fundamental picture of the world. Liberal values, modernity, and science were not positive notions here. In the yeshivah, people talked about evolution as a fiction—the Torah told the truth that the world was created 5,000 or so years ago. Being confronted with a completely different worldview threw my compass off. I felt dismayed, suspicious even, yet fascinated.

At one point, classes and lectures ended and everyone gathered for afternoon prayers. That's when I really got a jolt. Up until that point in my life, I had a fuzzy sense that I believed in God: a weak mix of the wonder of nature, an ordered universe, and some ethics thrown in. When I saw these men swaying their bodies, closing their eyes, whispering, shouting, and waving their arms in fervent prayer, I saw something more. They spoke directly to God. I thought: "I've been fooling myself! These guys really believe."

The world at Aish HaTorah seemed to me small, integrated, and crackling with spiritual energy. This world was only 5,000 years old, and the God who created it was right here listening to us! Watching us! That same God had revealed the divine will in the Torah given at Mount Sinai, and we Jews were the ones who needed to follow God's commandments. I stayed the night in their dorm. In the morning I even tried to change my plane ticket home, but the airline refused. I took that as a sign that my suspicions, my fears, were well-founded, so I threw my velvet yarmulke on the bed and tore out of there. Yet, this mix of attraction, suspicion, and yearning fed an inner turmoil that consumed me for many months afterward. I was caught in an either/or that I couldn't resolve.

It seems to me that I experienced a taste of something plaguing our world: the either/or schism between a powerful, integrated, communal, meaningful religious world and an alienated, shallow, lost modern world. Or is it a schism between an open, democratic, scientific, tolerant, progressive modern world and a fanatic, xenophobic, superstitious, backward, fundamentalist world? Take your pick.

I experienced conflict because each contains a part of the truth. And each has serious flaws. People are being pulled into extremist and fundamentalist religions—Jewish, Christian, Islamic, or other. We come to see these beliefs as the only antidote to the real problems of a world in which we live separate, desperately lonely lives. We are locked onto screens for company, working at jobs that don't offer any meaning, mesmerized by corporate-generated entertainment, fashion, news cycles, and politics that never touch our real humanity.

Some reject all religion as empty, as separating us from one another with old dogmas, as continuing centuries-old beliefs that may bring comfort to the weak-minded but are silly and unscientific. Religions are seen as dangerous, feeding war and division.

Which side is right? Of course, neither is right. If you and I and so many others like us hope to live an integrated life, we need to re-configure our categories so that we're not faced with either/or choices. We need a way to look at a paradox like this and hold on to both sides.

— N. M.

Introduction
Taking Apart or *Taking Part*

"When Ben Azzai, a student of the rabbis, was delving into the sacred writings almost 2,000 years ago, fire appeared all around him. Not normal fire, but flames that seemed to emanate from him. He wasn't burned, and in fact looked enraptured. The other students saw him and, thinking he was exploring some mystical practice that he wasn't ready for, called their teacher, Rabbi Akiba. When Rabbi Akiba came, he asked Ben Azzai, "Are you involved with the Chariot (the deepest, most secret mystical practice in those days)?" "No," he answered, "All I'm doing is stringing like pearls the words of the Torah together with the words of the Prophets and the words of the Prophets with the words of the Writings, and the words are as joyful as on the day they were given at Mount Sinai and as happy as when they were new." (adapted from Midrash Song of Songs Rabba 1:10)

This story, which inspired the title *The Pearl and the Flame*, is about an ancient Jewish way of thinking: Ben Azzai understands the holy writings by finding, or maybe creating, beautiful relationships between the words. He doesn't change one word of the sacred scriptures, but he reveals new meaning in them when he strings them together in new ways to create new juxtapositions and relationships.

He doesn't stand above the words to solve their contradictions or puzzles, but instead delves into them, partnering with them to gain a deep understanding that is joyful and creative. Apparently, the words also find it joyful! This way of thinking—looking for patterns and connections—is not only a way of exploring scriptures. It's a way of looking at the world. It was, and is, an

1

enlivening way for people to understand themselves and their place in the world.

We've learned to think differently in our times, not just about sacred texts but about many things. We've learned to buy our meat in plastic packages in a supermarket. We don't string together where that food came from. We don't weave together the pattern of who drove this meat in a truck and how many gallons of fossil fuel was used to get it here. We don't connect the strands of the story that would bring us into a relationship with the animal or animals who lost their lives so we may eat. We've created a fragmented world, breaking the necklace of stories that could connect us to our food systems, to our social and political systems and even to our own bodies. We've broken the lines of connection that give life joy and meaning.

How can we find wholeness in a world that is fragmenting? What are the ways forward in the face of the climate crisis, epidemics of addiction, and political and social upheaval? Can we reconnect to our spiritual traditions differently and at the same time harmonize these traditions with new trends in science? Could we use systems approaches, that is, those modern approaches to understanding such as Complex System Theory, Cybernetics, General Systems Theory, Emergence and more, that seek to explain phenomena not by taking them apart, but by looking at context, patterns and relationship? These questions all come down to the ways that we have learned to think, act and be in the world. That is where our work begins.

Many of us grew up in a world where serious thinking and work were associated with objectivity and linear problem solving. Thought is valued over emotions, and quite often, male is valued over female. When 17th-century philosophers such as Francis Bacon and René Descartes started looking at the world as a machine, it began a belief that the world could be broken down to its smallest parts, mastered and controlled by our manipulations. It brought into the world an intoxicating new sense of power and many advances, while at the same time causing a new alienation from the world.

Like the sorcerer's apprentice, we didn't realize that the new power in our hands would also engulf us. If we treated the world as a machine, we would come to treat ourselves as machines. If we perceived the world as so many isolated, individual things, we would come to see ourselves as isolated, individual things. We would blind ourselves to our connected relational nature.

We became alienated not only from our surroundings, from nature, but also from one another, from our own bodies, our very selves. Reducing everything to its smallest component is a powerful tool. It made us into powerful, god-like beings who learned to shape the world. This has brought many benefits to humanity. As I walk or jog through the woods every day with my dog, I can thank the miracles of modern science for my new titanium knees. I'm thankful that when my children fall ill, antibiotics are available. I'm thankful for hot showers in the morning. I'm thankful that I can board an airplane and visit my family on the other side of the world. So many things have been achieved through science and technology. Yet, the whole paradigm of reductionism on which our modern science and technology are built works by taking things apart, and finding the smallest units. Science has learned a lot, for example, about anatomy by doing dissections. But, as every high school biology student has learned, when we dissect a frog, it can no longer be alive. By breaking everything down to its smallest units we're basically dissecting our world—and it's dying.

For all our progress, this dissecting has fragmented our world and life-sustaining systems, both natural and social, are falling apart. We've altered the climate to the point that as I write this the Amazon is burning; droughts, floods, hurricanes are ravaging cities and villages; democracies are faltering as dictators and demagogues spring up in country after country. At some point, we must come to the realization that, to paraphrase Albert Einstein,

The *Pearl* and the *Flame*

"…we can't solve problems with the same type of thinking that we used when we created them."[1]

Traditional cultures, indigenous peoples, and ancient traditions—such as Judaism, the one I know intimately—have always looked at the world differently. There is an organic way of thinking that harmonizes with nature and with our human nature. For three centuries, at least, the Western world has taken a different path, and the ancient wisdom has been exiled, ridiculed, and misunderstood. Judaism has tried to fit in, tried to cut out the embarrassing, "illogical," "primitive" parts, and succeeded mostly in becoming politely irrelevant.

Yet, in recent times we have seen a flowering of new perspectives: ecological thinking, cybernetics, systems thinking, complexity, and other forms of relationship-based thought. Western knowledge (or at least some part of it) is starting to shift toward more holistic paradigms in greater harmony with the ancient traditions. Instead of the ruling paradigm of explaining through reducing phenomena to an underlying cause (appropriately called "reductionism"), these new paradigms look for an explanation in relationships, context, patterns, and systems.

With these new, non-reductionist paradigms, modern scientific knowledge need not reject what is ancient but may benefit from the power of these traditions. Similarly, rather than breaking with modernity in a violent reaction of extremist religion, religious traditions can come together with modern science in mutual understanding of a new paradigm that embraces both, offering a coherent approach to living in an era of rapid change and confusion.

---

1 Apparently Einstein said something similar to this. One possibility is "The world we have made as a result of the level of thinking we have done thus far creates problems that we cannot solve at the level of thinking at which we created them." The phrasing used here has become popular, however, and its likely origin is with Ram Das from the early 1970s, loosely paraphrasing Einstein.

4

The "Three Mems"

This book will focus on three modern, non-reductionist concepts from the world of systems thinking. These are Emergence, Nestedness, and Bifurcation events, or "tipping points." These concepts will be mapped onto corresponding core concepts in Judaism, which can be expressed in three Hebrew words all beginning with the letter Mem. These are: *Minyan* (a quorum of ten, which is an example of a whole greater than the sum of its parts), *Mikdash* (sanctuary, which is an example of a nested structure in which holiness is contained in nested patterns) and *Mitzvah* (an example of a small action which can have huge consequences, though this is never predictable: tipping points). I call them the *"Three Mems"* and this book is organized around these three correspondences.[2]

A Broader View

A story often told by environmentalists illustrates the shortcomings of narrow, linear thinking.

> Many Dayak villagers had malaria, and the World Health Organization had a solution that was simple and direct. Spraying DDT seemed to work: mosquitoes died, and malaria declined. But then an expanding web of side effects ("consequences you didn't think of," quips biologist Garrett Hardin, "the existence of which you will deny as long as possible") started to appear. The roofs of people's houses began to collapse, because the DDT had also killed tiny parasitic wasps that had previously controlled thatch-eating caterpillars. The colonial government

---

2 I was influenced to use these terms after reading Tom Wessels' clear and concise description of systems thinking in the context of ecology in his book *The Myth of Progress: Toward a Sustainable Future* (Burlington, University of Vermont Press), 2000.

issued sheet-metal replacement roofs, but people couldn't sleep when tropical rains turned the tin roofs into drums. Meanwhile, the DDT-poisoned bugs were being eaten by geckos, which were being eaten by cats. The DDT invisibly built up in the food chain and began to kill the cats. Without the cats, the rats multiplied. The World Health Organization, threatened by potential outbreaks of typhus and sylvatic plague, which it had itself created, was obligated to parachute fourteen thousand live cats into Borneo. Thus occurred Operation Cat Drop, one of the odder missions of the British Royal Air Force.[3]

Current examples of this sort of thinking abound:

Burning fossil fuels put so much carbon into the atmosphere that the world's climate is changing, melting the polar ice caps. How are governments and oil companies responding? With a rush to exploit new sources of Arctic oil that were previously impossible to reach because of ice. That makes sense only if we ignore the broader context and focus on short-term profit. It sounds good only to a mind that doesn't consider the whole picture.

Refugees gather on our southern border in growing numbers. Putting up a wall and detaining children in inhumane conditions are offered as ways to solve the problem, but it would be better to look at the whole system. People flee their homes in Central America for many reasons, including economic devastation caused by U.S. trade agreements such as NAFTA.[4] Another major reason is an attempt to escape the dangerous drug wars there, and the cartels and gangs responsible for the lawlessness depend on the demand for illegal drugs here in the United States.

---

3 Amory Lovins, Hunter Lovins, Paul Hawken, *Natural Capitalism: Creating the Next Industrial Revolution* (Little, Brown and Company) 1999, 285-86.
4 See for example: https://www.commondreams.org/views

The U.S. is experiencing its own epidemic of gun violence, to the extent that some people are afraid to go to work or send their kids to school. One way to feel safer is to carry a gun, but will more guns on the street help the situation? That might feel true from the individual perspective—a feeling reinforced by the powerful gun lobby—but from the systems perspective, one person getting a gun leads to the next person feeling the need to be armed. So, gun ownership multiplies, increasing the incidence of homicides, suicides, and accidental shootings. As is often pointed out: If guns made us safer, we'd be the safest country on earth.

One basic fact underlies these examples and others that could be named: Our reductionist, atomistic perspective—seeing ourselves as isolated, free-floating individuals instead of as parts of a greater whole—has led to terrible harm to the living systems of the world, both biological and social. It has caused us to be more depressed and more anxious; to need more drugs, more consumer products, more on-screen distractions; to turn in fear to the politics of isolationism and racism.

Perceiving isolated individuals and events without looking at the network of relationships that bring them into being, has brought about environmental disaster, political chaos, and psychological suffering. This contextless orientation creates a feedback loop of increasing isolation, distrust, distraction and disease, and seems natural to us because it is the environment in which we were raised. It is the ocean that we swim in. In order to start to climb out of that ocean to get a glimpse of something else, let me start with my story, which began on an island.

Complicated Islands: Where I Came From

I grew up in Honolulu, Hawaii, across the road from a bay framed on the East and West by the shapely forms of extinct volcanoes: Koko Head, Koko Crater (*Kohelepelepe*), and Diamond Head (*Le'ahi*) crater. We got the back side of Diamond Head, not the iconic view from Waikiki that you see on postcards, but it was still beautiful. Rainbows appeared just about every day as the misty rain filled the backs of the valleys but left the sky over the beaches sunny and blue. Our home was tucked against the mountain, surrounded by huge banyan trees whose vines hung from muscular, curving branches. These vines attached themselves to the ground, put down roots, and thickened into new trunks, forming a complex arboreal architecture of intertwining caverns, bridges, and cathedrals. My two sisters, my brother, and I loved to climb and play in those trunks, branches, and vines.

Of course, it wasn't all paradise. I was the son of transplanted New York Jewish intellectuals in what was at the time quite a provincial, conformist, culturally conservative social environment. Our house was dark, set back from the street and under those massive banyans. I imagine that our neighbors saw us as strange, awkward, out-of-touch people in a ramshackle, perhaps haunted, house.

It seemed to me that we stood out in sharp contrast to the tidy homes with immaculate lawns that surrounded us. Those homes held normal people: They were tanned, pidgin-English-speaking, mostly of East Asian origin[5], or some version of the local ethnic mix of Hawaiian, Portuguese, Filipino, and a smattering of Haoles (Caucasians). They all seemed in the know, cooler, more connected to the local scene.

My best friend in elementary school was Greg Kageyama. We used to play touch football after school with all the neighbors on his street. We seldom came over to my house. His father owned a franchise of Midas Mufflers. I knew of almost no other parents who were, like my dad, professors; mostly they were business people like Greg's dad. "Jewish" wasn't really even a category in my neighborhood. Maybe a slightly more obscure version of Haole, which already stood at the bottom of the coolness scale.

---

5 Growing up in Hawaii I never heard the term "Asian" as an ethnic description, as it is commonly used on the Mainland. Most people were some form of Asian so it wasn't a relevant or useful category. You were either Japanese, Chinese, Filipino, Hawaiian, etc, or, quite often, a mixture, known as "Hapa" in Hawaiian.

So our home was both a refuge and a prison. Those big banyan trees surrounding our house became for me almost like alter-egos. I loved their beauty, power, and sensual patterns. I loved their shade and the possibilities for games of hide and seek that their roots and vines provided. But they also stood for how we were different; cut off and separated from our attractively normal neighbors. There was love, pride, and also sadness, frustration, and embarrassment held in the shadow of those trees.

The inside of my house also mirrored those feelings. The walls were dark, textured wood panels; lined with classic old books. Much of the furniture was from my parents' home apartments in New York: dark wood and leather, whispering tales of Old-World gravitas and mustiness.

Paintings hung on the walls, almost all done by my father's father, Joseph Margulies, who was an accomplished New York artist. Never reaching the highest levels of fame or riches, he nevertheless was well known and exhibited in important galleries and museum collections. His bread and butter living came through painting portraits of wealthy Jewish families, in the decades of the 20th century when having an original portrait was a status symbol, a sign of having made it in America. To this day there hangs in that house a painting of my father at age four or five and next to it one of a little girl who would, one day, be my mother, but who at the time was the young daughter of an up-and-coming ophthalmologist, Abraham Kornzwieg, and his wife Chifra.

In addition to those portraits, my grandfather would also, so they told me, go down to the Lower East Side and give the old, religious Jews little bribes to let him sketch them as they studied Torah or smoked their pipes and schmoozed. So, we also had several drawings and etchings of wise-looking old men with long beards and yarmulkes peering into books, or staring out of the picture straight at us, as if challenging us moderns to justify our new ways of life: "Are your new ways any better than what I've got?"

The *Pearl* and the *Flame*

These weren't the kitschy, mournfully praying rabbis or dancing Hasids that you see in Judaica shops. My grandfather, who was secular, called them "characters." By that he meant he saw something special and interesting in these traditional people: they were his people, yet they inhabited another world, and he wanted to capture it in his art. Despite having to paint those portraits of the newly rich American New York Jews, he was a real artist, and he knew how to capture character with a few lines.

Those old bearded men, like the banyan trees, found their way into my soul. They represented something that I knew was in my family but further back—something about which I didn't have much of a clue. I had a sense of pride and joy in our family's intellectual life as I listened to my parents' discussions over the dinner table. I also felt a kind of security in what was clearly the serious, dedicated devotion our parents had to one another and to us kids. We were taught to be good and considerate, to do well in school, and to find ways to serve humanity. I associated these values loosely with our Jewishness, but the connection wasn't very clear, and they were mixed in with all that made us uncool, un-local nerds.

Meanwhile, those trees beckoned to me, as did the lush land and the blue, wild ocean . I yearned for the wild, where all this complicated social world where I didn't fit in would disappear and life would abound in vitality. When I was five or six years old I loved to have my father read to me the Jack London classic *The Call of the Wild*, in which a domesticated dog, Buck, is taken from his soft life in California to the harsh frozen trails of the huskies in the Alaskan gold rush. Ancient, wild instincts well up in Buck. He struggles, fights and survives to find in himself, by the end of the book, the "dominant, primordial beast."[6] In the final pages he runs with joy at the head of the wolf pack, free of the fetters of humanity. I had the book practically memorized.

---

6 Page 61 in my grandparent's 1913 Grosset and Dunlap Publishers edition, now on my shelf.

When I was in seventh grade, my social studies teacher, Mr. Yamamoto, gave me the book *The Naked Ape* by Desmond Morris, which was a popularization of research into how human behavior is influenced by our primate origins, what is called today evolutionary psychology. It set me to several years of thinking that I would save humanity from itself by uncovering the instinctual, real human ways of being that lay buried in our ancient, primate genes. I was still searching for Buck.

These and many other youthful searches came and went. I was searching for the vitality of life, and I desperately wanted to feel a part of, not apart from, it. But I didn't know quite what I wanted to be a part of or how to get there. Was it being "local," surfing and talking pidgin? Was it to be found in these books and intellectual theories connecting me back to something real, natural, and alive? All I knew was that the conventional life I was living wasn't enough for me. For me to begin to "take part" instead of taking apart, I would need to break free from much of the modern life that I knew.

PART ONE:

Our Relation to Creation

Ancient Wisdom in a Complex World
Stringing Pearls

Looking back, it's not surprising that in my early 20s I was off looking for my soul's deepest roots across the world, in Israel. There, of all the places I had searched, I finally felt I had found something I could be a part of. It was something about the land and the people, but also the yeshivah, the study house, where I learned Jewish texts such as the *Talmud* and *Mishnah* and *Midrash*. How could my Hawaiian-Jewish soul find its deepest connection in ancient Jewish texts?

It came back to finding a way of thinking and being that starts with relationships. I began to see that in these Jewish texts, and in Jewish life, understanding comes from putting things together, from finding relationships and building patterns. It is reminiscent of those winding roots and vines of the banyan tree which felt so vital and unfettered, beautiful in their intricate patterns. In these texts I found a way of thinking that felt both intellectually exciting to me and also partaking of that wild, vital, thread of life that I sought.

This flexible, creative way of thinking connects us to the world instead of placing ourselves above it. So, let me return now to the story that I opened with: Ben Azzai and stringing words like pearls and the flames surrounding him. Let's dig a bit deeper into this story.

Ben Azzai was a student of the great 2nd-century sage Rabbi Akiba, and he was an intense guy. Saying he was dedicated to his studies would be a gross understatement. Once, when his colleagues pointed out that he was not married, even though he knew very well that marriage and bringing children into the world was a central value of the Torah, he replied bluntly: "What can I do? My soul yearns for Torah! The world will have to be populated by others." (Babylonian Talmud Yevamot, 63b)

What do we know about this intense love for the Torah? What was it about learning Torah that aroused such intense passion for him? The story I told earlier about him is actually a midrash. "Midrash" is a rabbinic way of "delving" or "exploring" some biblical text. *Midrashim* (the plural of *midrash*) often come in the form of stories. That means the story of Ben Azzai doesn't stand alone but came about as a commentary on a verse from the Scriptures. It makes sense that a story about making connections wouldn't stand alone, but would already participate in the connections it expounds on.

The verse that the rabbis are exploring is from the Song of Songs, a dream-like love poem which almost didn't make it into the biblical canon. Ben Azzai's teacher, Rabbi Akiba, came to its defense, saying that not only should the Song of Songs be included in the biblical canon, but that if all the other writings are holy, this is the Holy of Holies. (Mishnah Yadayim 3:5)

Beginning the poem, we hear lines like these: *"Oh, give me of the kisses of your mouth, for your love is more delightful than wine. . . . Your ointments yield a sweet fragrance, your name is like finest oil – Therefore do maidens love you. . . . Draw me after you, let us run! The king has brought me to his chambers. Let us delight and rejoice in your love, savoring it more than wine – like new wine they love you!* (Song of Songs 1:1-4)

A few lines later it says: *"Your cheeks are comely with plaited wreaths, your neck with strings of pearls."* The rabbis who created this midrash pick up on that phrase "strings of pearls." The rabbinic imagination hears the words "your neck (is lovely) with a string of pearls" and decides to tell the story of Ben Azzai. Here's a more direct translation of the original text:

Ben Azzai was sitting and learning and there was fire all around him. The other students saw and, thinking he was involved in the mystical practice, the Chambers of the Chariot, which would be forbidden for such a young student, called their teacher, Rabbi Akiba. When Rabbi Akiba came, he asked Ben Azzai, "Are you involved in the Chamber of Chariots?" "No," he answered, "All I'm doing is stringing the words of the Torah together with the words of the Prophets and the words of the Prophets with the words of the Writings, and the words are as joyful as on the day they were given at Mount Sinai and as happy as when they were new. As it says (Deut. 4:11), 'The mountain burned with a fire...'"

Let's unpack this. In order to understand the verse "your neck with strings of pearls" the rabbis tell a story about one of their own, about themselves, stringing words. The words of Torah are, apparently, like pearls or jewels. For the rabbis, and perhaps especially for Ben Azzai, the act of learning Torah is an act of love. The fire that surrounds Ben Azzai is a fire of passion as he communes with the Torah. But as we see from the verse that he quotes referencing the revelation on Mount Sinai, it is also the fire of revelation. What is it about the simple act of stringing words together that feels like a revelation?

It may seem simple, but this way of understanding is radical: In order to understand a verse in the Torah, one doesn't isolate it and try to find "the answer" to its true meaning. One doesn't get the meaning of a word by looking it up in the dictionary. Instead, one enters a creative process by looking for relationships and juxtapositions. With every new context, pairing one text with another, new meanings emerge. In the process of *midrash* (delving, searching), the ancient rabbis partnered with the Torah, intimately co-creating an infinitely fruitful source of revelation.

This rabbinic activity, called *midrash*, of stringing words from different contexts together; creating new combinations that were akin to new revelations, was only the more radical aspect of something much older: telling stories. Stories connect us to life through stringing together events, places, and people with words. Telling stories is essential to being human. (Gregory Bateson quipped in his 1979 book *Mind and Nature* that we'll know that computers have reached a human kind of intelligence when we ask the computer, "Can you think like a human?" and the computer answers, "That reminds me of a story.")

Indigenous cultures always start with stories. The writer Martin Prechtel uses the same analogy to stringing a jeweled necklace:

> Our soul can only understand the world as a story, a mythology in which the things of the past and the random happenings of our present are strung together like beads on a string of continual story in which reality is remembered as a sequence of jeweled events. Like a bead on a rosary of nature's moods, each thing that happens is relevant to the beauty of the entire necklace. This necklace begins way before us and we are strung onto it, and in this way the Holy is adorned by our participation in this necklace of mythic sequence from the past to our present and beyond. Hence everyday lives become relevant to the maintenance of that ongoing storied beauty as the Holy keeps on stringing life together beyond us.[7]

Humans need meaning in our lives, and telling stories brings us that meaning. The radical thing that the ancient rabbinic sages so loved was creating new stories without throwing out the old ones. They were able to keep the Torah alive and relevant in a changing world because they found ways to be completely inside the Torah, to delve deeply into its meaning but also to find new meanings there—meanings that hadn't been seen before because no one had ever made that juxtaposition before. Or maybe that juxtaposition was always meant to be found and was there from the day it was given on Mount Sinai, waiting in potential. The

---

7 Martin Prechtel, *The Unlikely Peace of Cuchumaquic: The Parallel Lives of People as Plants: Keeping the Seeds Alive* (Berkeley, North Atlantic Books) 2011, 401.

Torah comes alive when a person dives into her words; they are joyful, reliving the day they were first revealed, and the person is on fire with love.

Where did this way of thinking come from? How did ancient rabbis such as Ben Azzai come upon a way of thinking that emphasized relationship and connection? The Torah, the texts and the stories and ways of thinking about the world had been passed down to them, and they knew that merely transmitting the outer shell, the words, wouldn't be enough. Nor was it enough simply to tell the stories. As we saw with Ben Azzai, his goal was something like a new revelation. The goal was to transmit a way of thinking, something that existed between the lines. That way of thinking had its origin in the thousand years that their ancestors had lived on the land, farming it and learning its ways.

Cultures of Relationship
& Cultures of Control

The hilly boondocks of Canaan were not good for the big, industrial agriculture that made Egypt and Mesopotamia the superpowers of their day. Those regions with their various empires grew up around great river basins. Their material wealth was a result of organizing many slaves and serfs to work the irrigated fields and the technology of canals and pumps to move the regular rhythm of the rivers.

No, Canaan was a country dependent on the rain of heaven falling in its season. When rain fell they knew it was a blessing from God. It was a land where the people needed to improvise, to care about the details of their very local ecosystem. One side of a hill was not going to get the same rainfall as the other side. They needed to maintain good relations with their neighbors, who would help with the tasks of small farming and raising flocks, and they needed to respect the land if it was going to sustain them and their family for generations. They needed to create the kind of society that God would want to bless.

So, the way of life and religion of Ancient Israel was very much a counter-culture and an alternative to the big powers. It was a way of life and religion (the two weren't separate back then) that depended on relationships with the whole network of natural and human and divine partners. Within those relationships the fruits of the earth, the fruits of one's labors, were seen as gifts to be valued, as all gifts should be, not just for themselves but as outward signs of a relationship.

If I were a farmer in those days, I'd bring my first fruits up to the Temple in Jerusalem to present them to the *kohen* (priest). It was a way of showing that I valued the land and its fertility as a sign of my relationship with the Source of All. And that relationship was not just with me, the farmer; when I presented that basket of fruit at the Temple, I recited the whole story of my people. In fact, the core of the Passover *Haggadah*, still recited every year by Jewish families around the world, is derived from this "first fruits" recitation: "My father was a wandering Aramean, and he went down to Egypt with meager numbers ... but God brought him up with an outstretched arm...." (Deut. 26). Beyond my personal life history, I was a part of a centuries-old relationship between the Israelite people and God. My relationships were not just about the present moment but stretched into the past and future as well. The stories that these Israelites told were the connecting threads that held them together as a people, and they also maintained their proper relationship with the land, with Creation.

Rain vs. River:
An Ecological and Literary Pattern

To go deeper into the contrast between ancient Canaan and Egypt or Mesopotamia, we can look at a pattern in the biblical text that points us to an awareness of ecology as a factor in their consciousness.[8] Start with this passage from Deuteronomy 11:10-12:

> For the land that you are about to enter and possess is not like the land of Egypt from which you have come. There the grain you sowed had to be watered by your own labors (literally: was watered by your foot), like a vegetable garden; but the land you are about to cross into and possess, a land of hills and valleys,

---

8 See, in this regard, the excellent book *The Ecology of Eden,* (New York, Vintage Books), 1999, by Evan Eisenberg. Eisenberg discusses this contrast between river irrigation and rain dependent economies and I am indebted to him for his insights.

> soaks up its water from the rains of heaven. It is a
> land which the LORD your God looks after, on which
> the LORD your God always keeps His eye, from the
> year's beginning to year's end.

These verses are part of Moses' speech to the Israelites on the brink of entering the land of Canaan. They are phrased as a praise of this new, "promised land," along the lines of "a land flowing with milk and honey." It'll be great! Rain just comes down from heaven! Less work! But it is more complex than that. We can't ignore the next line that talks about God's eyes on the land from the beginning of the year to the end. Sounds a little scary.

And what is going on with Egypt? The JPS translation says "There the grain you sowed had to be watered by your own labors ..." but that "had to be" isn't in the words, it's the interpretation of the translator. The Hebrew text just says "where you watered your grain with your foot." It's a reference to the irrigation system where they might have had channels and pumps that were operated by their feet. It was a system that depended on the regular overflowing of the Nile and the work of a hierarchical, mass labor system that harnessed the bounty of the river into a machine of production, regular as clockwork; they never needed to look up to wonder whether there would be a sustaining, nourishing rain. Their focus was down on their feet.

A hint of what is being invoked here can be seen in the story of the serpent in the garden of Eden. Strange as it may seem, at the start of the story the serpent has legs. We know that because it is only at the end of the story that we're told that the serpent will now crawl on his belly and literally eat the dust for his food "on your belly shall you crawl. And dirt shall you eat all the days of your life." (Gen. 3:14) The Hasidic master Rebbe Nachman of Breslav focuses on this downward stare of the serpent and comments that the dirt (being the most material thing that one can imagine) represents materiality and money. Playing with the syntax, he reads the verse to say that the obsession with materiality/money eats up all the days of a person's life. (Likutei Moharan 23:6:7)

The Egyptians in this comparison are like the serpent. They are caught in an illusion: that they can engineer their way to abundance and security. If they control and manipulate skillfully enough, they can always count on a steady, stable food supply. The problem is that this isn't how life works. Change is inevitable. As the saying goes, "sh-t happens." Or, more positively, as they learned the hard way in *Jurassic Park*, "Life finds a way."

The Egyptians (or at least the Egyptians in the imagination of the Ancient Hebrew writers who composed these stories), with their pyramids and priestly bureaucrats, had bought into the seductive illusion that humans can use our engineering and technology to control our way to perfect convenience and safety. The elaborate tombs and mummification techniques were attempts to conquer the final enemy: death. Paradoxically this obsessive need for ultimate power and control that tried to deny death squeezes out all tolerance for the dynamic, unpredictable miracle called life.

This is why the fertility of the Israelites described in the first chapter of Exodus was such a thorn in the Egyptians' sides. "The more they were oppressed, the more they increased and spread out. So that the Egyptians came to dread (literally felt "thorned by") the Israelites (Exodus 1:12). The rabbinic midrash says that they had six children for every birth, other opinions say 60, and other opinions say 600! The hyperbole of the midrash is even more pointed when we consider the reality of slavery as we know it from modern times. Ilana Pardes, in her book, *The Biography of Ancient Israel.* Points out that people who are caught in the inhuman suffering of being enslaved in fact have very low fertility.[9]

---

9 Ilana Pardes, *The Biography of Ancient Israel: National Narratives in the Bible* (Berkeley, University of California Press) 2000, 22.

This hyper-fertility of the Israelites, thus, was not natural, but can be seen as a divine response (a literary divine response; I'm not making any claims here about what "really happened") to Egyptian attempts to control and enslave the Israelites and by extension, the whole Egyptian project of controlling, taming and capturing resources, nature, and life. The Egyptian's exaggerated fear and loathing of the Hebrews can also be seen as reflecting the Egyptians' own xenophobic response to their foreign workers. These aliens were about to sap all their resources with their prodigious birth rates! They were about to poison the body politic with their foreign and un-Egyptian ways! We know this storyline only too well today, as it is used by those who want to whip up fear of the foreigner, the Other.

The biblical text takes another swipe at this Egyptian xenophobia a few verses later when it describes how the midwives fool Pharaoh by using his own racist prejudices against him.[10] Pharaoh had ordered the midwives to kill all the male children as soon as they were born. But the midwives, being God-fearing women, didn't do so. When Pharaoh called them in and demanded to know why not, they told him, "because the Hebrew women are not like the Egyptian women: they are vigorous. Before the midwife can come to them, they have given birth." (Exodus 1:19) The word translated as "vigorous" can also mean "they are animals." This, apparently, was a reasonable explanation in Pharaoh's mind, so he accepted it and went on to Plan B: throwing the male babies into the Nile

This opposition of the Egyptians to the fertility of the Israelites is succinctly and shockingly portrayed in another midrash told by the rabbis in which they picture God taking counsel with the Heavenly court about whether to intervene on the side of the Israelite slaves in Egypt. In the middle of the debate, the angel Gabriel approaches the Divine throne holding a brick used in the building of the pyramids. As the angel gets closer to the throne,

---

10 *The Biography of Ancient Israel*, 20.

the gathered angels and divine beings can see that baked into the brick is an Israelite baby. That is all that God needs to see, and God comes down to save them.

The messy yet joyous phenomenon of having and raising children stands in opposition to the cold, manipulative technology of the Egyptian culture. This is one aspect of the contrast that our verse in Deuteronomy is talking about when it compares the hills and valleys watered by rain with the Egyptian fields watered by "your feet." The Israelite economy, based on small farming on difficult terrain, was all about living a life of dynamic relationships: with the land, with the flora and fauna, with neighbors and with God. As in all relationships, you need to listen, respond to feedback—you're not in complete control, but if you're diligent and sensitive you can keep the relationship flourishing.

I'm reminded of our older son Nadav's bar mitzvah *D'var Torah* (speech or sermon) a couple of years ago, which has kind of found a place in synagogue lore as one of the more impressively shocking bar mitzvah speeches. The back-story: We had started out sending our kids to a pluralistic Jewish day school which we loved, but ultimately didn't fit our kids' personalities, learning styles and—when it came down to it—religious beliefs. Nadav was never drawn to the spiritual and religious teachings that were offered there. It wasn't his thing. Yet, we persisted in keeping him in that school until the sixth grade before we finally took him out.

His bar mitzvah portion was Va'era, the section of Genesis which includes the harrowing story of the Binding of Isaac. That is where Abraham, the Patriarch of the Jewish people and the father of Isaac, responds to God's test—the command that he sacrifice his beloved son—by diligently obeying. Isaac is only saved when, at the last second, an angel calls out "Abraham, Abraham! Do not raise your hand to the boy or do anything to him." (Gen. 22:11-12).

Nadav was clear all along that his thesis would be that Abraham failed the test. He should have stood up for what was clearly right and refused God's outlandish and cruel command. So far so good—that was a reasonable interpretation which expressed a strong moral sense and a healthy questioning of authority. Where he really got himself into the annals of shul history was when he compared us, his parents, to Abraham—sacrificing him on the altar of our misguided religious ideas by sending him to Jewish day school—until we finally heard the angel's voice and mercifully took him out. (He did allow that we were good parents for at least belatedly listening to that voice of reason). He introduced these ideas with the memorable line "My parents believe in God and are devoted to Judaism. Me.... not so much."

Thankfully, we are good parents and we go to a good synagogue. We, along with the rabbi and the rest of the congregation could not be prouder of him and his holy chutzpah in expressing his true beliefs. Is he on the religious path that we had hoped for when we started out sending him to Jewish day school? Not exactly. But supporting him in finding his own voice is our main job as parents and we believe it's ultimately the best way to approach his Jewish education. Raising children is messy because each one is unique and they put us in our places when we try to impose our paths too uniformly on them. Life finds a way, or to use a Jewish phrase: Choose Life! That means listening and responding to the ever-changing, challenging and delightful surprises of our children and all of our relationships.

Coming back to those thin soiled, difficult to farm, hills and valleys of ancient Canaan: the ecology and economy was completely dependent on rain. It "soaks up its water from the rains of heaven." When the rain fell, it was a sign that their relationship with God was good. The rain was a blessing from God. When the rain didn't fall, it was a sign that one needed to examine what might have gone wrong.

The very next paragraph, which is very familiar to observant Jews because it is recited daily as the second paragraph of the Sh'ma, explains the relationship: "If you will surely follow my commandments then I will bring the rain in its season...." The rain is seen as a gift from God and, like all gifts, there is the object itself, but the true value is in the relationship that it expresses. It can sometimes be difficult for us to understand the importance of gifts because we've relegated them in our minds at least to trivial extras—our main economy is all about commodities. But for many societies, especially traditional societies, the main economy is a gift economy. Property travels from one person or group to another, but the real currency is the set of relationships that are established and maintained which keeps the current of gifts flowing.[11]

I remember trying to explain this idea, that the most important part of a gift is the relationship, to Nadav after his bar mitzvah. To tell the truth, he was pretty into the money and other gifts he received. But from the parents' point of view, it's clear that friends and family are giving him bar mitzvah gifts because they are affirming their relationship with us as a family. From our slightly wider perspective, these relationships are much more important than the actual money. When the gift-givers hold their own bar or bat mitzvah or bris or marriage, we will do the same for them. The gifts are primarily affirmations of the relationships that make up our community. Nadav may have focused more on the $36 or $72 checks that he received (it's a tradition to give a multiple of 18, which in Hebrew numerology is "Life"—and it now occurs to me that perhaps that tradition is a subtle way of reminding ourselves not to get caught up in the monetary value, but to remember the vital life of the gift), but we know that over the years these relationships are what make our marriage, our

---

11 See Lewis Hyde's classic book *The Gift: Imagination and the Erotic Life of Property* (New York, Vintage Books), 1979. He puts "Imagination" in the subtitle because artistic creation always comes to the artist as a gift, a mysterious grace that follows hard work and discipline, but is magically beyond it. "Erotic Life of Property" is meant not in the sexual sense, but invokes property as that which travels between people and groups and creates connections and relationships. See also Robin Wall Kimmerer's beautiful personal description of her gift relationship with strawberries and other plants in *Braiding Sweetgrass: Indigenous Wisdom, Scientific Knowledge, and the Teachings of Plants* (Canada, Milkweed Editions), 2013, 22-32.

family and our community strong and thriving.

In ancient Canaan, the small farmers and pastoralists needed the rain in its season. They also needed to understand the land and care for it. The mountainous terrain made for many small micro-climates. The importance of these micro-climates is captured in biblical ritual. The biblical account says that Joshua recited blessings and curses for the nation on Mount Gerizim and Mount Ebal (Deut. 11:26-30, Joshua 8:30-35). The blessings were shouted from Mount Gerizim and the curses were shouted from Mount Ebal. If you go to those places today, near the city of Nablus, or Shechem, you'll see that Mount Gerizim is on the windward side and is green and lush, while Mount Eval is on the leeward side and is brown and bare. The farmers also needed to cooperate and help one another through droughts, storms and all the vicissitudes of life on marginal, demanding land. Lending to one another (interest free), returning lost objects or animals, helping to unburden a struggling animal, or inviting guests in with generosity are all frequent themes of the biblical text, (and foundational parts of Jewish law to this day).

In short, relationships were key to the ongoing survival and flourishing of the ancient Israelites. The hilly, thin-soiled land demanded that kind of small-scale, relationship-based economy, and the religion reinforced those values. Egypt was a primary foil, standing for the opposite, but it wasn't the only example. When we think of negative places in the biblical stories, a few of them come to mind: Sodom and Gomorrah, for example. Lot, Abraham's nephew, wanted to split off from his uncle and start out on his own. Abraham said he could go anywhere he wanted. This is how the text describes Lot's response:

> Lot looked about him and saw how well watered was the whole plain of the Jordan, all of it – this was before the LORD had destroyed Sodom and Gomorrah – all the way to Zoar, like the garden of the LORD, like the land of Egypt. So, Lot chose for himself the whole

plain of the Jordan, and Lot journeyed Eastward. (Genesis 13:10-11, JPS)

We know how that turned out. The text even compares the Jordan River valley to Egypt. We know from the story of Lot's experiences in Sodom that their defining characteristic was their absolute prohibition against showing hospitality to guests. They were a rich, "gated community" – the date palm capital of the Middle East, with their income guaranteed because of the flowing river that watered their palm trees. Yet, that "guarantee" didn't make them happy and generous; it only made them more obsessed with keeping all their wealth to themselves.

Where else do we find the river-valley, lock-in-our-wealth-and-security syndrome? Not far from the story of Lot and Sodom is the earlier story of the Tower of Babel. This is actually a parody of the other major civilization that the Israelites saw themselves in opposition to: the Mesopotamian civilizations of the, you guessed it, river valleys of the Tigris and Euphrates rivers. This was the famous Fertile Crescent, one of the places where agriculture started, and the place Abraham left to start his spiritual journey to the backcountry of Canaan.

> Everyone on earth had the same language and the same words. And, as they migrated from the east, they came upon a valley in the land of Shinar and settled there. They said to one another, Come, let us make bricks and burn them hard. Brick served them as stone and bitumen served them as mortar. And they said, "Come, let us build us a city, and a tower with its top in the sky, to make a name for ourselves; else we shall be scattered all over the world. (Genesis 11:1-4, JPS)

Never mind that only two chapters back, the survivors of the flood had been commanded to "fill the earth," not to gather in one valley. Their "one language and one speech" can be seen as a hint toward totalitarian rule: one goal, one opinion, one idea: build the tower. The rabbis in the midrash (Genesis Rabbah 37:4; quoted in Rashi on Genesis 6:17) even imagine that the name of the place that they settle, "Shinar," is a hint to all the bones of those killed in the flood when it drained down to this one low sinkhole, where they *sh'nina'er* —"are shaken down"—as in a whirlpool at the bottom of a sink. So, the great Tower that is supposed to reach the sky is actually built, symbolically, on a low point, on the fear of the flood. The waters of the flood that destroyed everything became their unconscious enemy, and they would stop at nothing to build their way to solidity, solidarity, and security.[12]

All these river cultures, as they are represented in the biblical texts, repudiate the dynamics of relationships in favor of building a tower, engineering a river, or locking out strangers, in other words, trying to create a bubble that will shield them from change. They are convinced that they can control and manipulate the world so that they never need worry. They can conquer life and all its contingencies and create a fail-safe, airtight world. It never works.

The Origin of Waste

These cultures (or at least the caricature of them that is created in the biblical account) created the conditions for waste. Waste doesn't exist in a healthy natural system. Everything is used. Death and decay are inevitable and difficult, sometimes tragic, for the individual, but they are not waste; they are necessary parts of the cycle and keep the system going. The beauty of a forest or any natural system is as much about death as life, as the hardened, twisted and bruised trunks of trees and the dead trees on the ground all come together into what to our eyes is a vibrant, life-filled natural scene.

---

12 I first heard this explanation in a class given by Avivah Zornberg in a class in Jerusalem.

But, for those cultures that have become obsessed with control and with their power to keep all the contingencies and accidents of life at bay, they create a new, unnatural, category: "waste." There are those things that they want to keep in, and then there are the unruly other parts of life that are out. The children of the Children of Israel were a thorn in the sides of the Egyptians. The people of Sodom were unwelcoming of guests. The builders of the Tower of Babel were intolerant of divergent opinions or plans.

We have, in our modern world, largely adopted this culture of control. For a long time, the world seemed so big that it didn't matter that we put our waste outside, away, but we are now coming to realize that there is no "away." On a planet connected by wind-streams, oceans and moving animals there are no real borders. Yet we continue in our mindset of "away." If we don't wake up to our illusion, we risk following the examples of Egypt, Sodom and Gomorrah, and the Tower of Babel: seeing our "control" unravel into chaos. We see it happening already. The Israelite religion tried to point to a different way: the way of relationship, dynamic interactions in an ever-changing flow of life that could be hard but also blessed and joyful.[13]

The Rabbinical Alchemy: Life-Containing Texts

It was this relational religion and way of life that the rabbis, after the Romans destroyed the second Temple in 70 C.E. and exiled the Jews from their land, managed to transform into texts. These texts started out as oral discussions, questions, arguments, more questions. Over the course of a few centuries these discussions and arguments were remembered, recorded, and edited, and they became the Midrash, Mishnah and Talmud: the foundation of Jewish textual culture. It was as if they needed to

---

13 For an expanded version of this thesis see my article "From Waste to Wonder: Steps to a Spiritual Ecology of Living," *Tikkun Magazine*, Vol. 19, no. 4, July/ August, 2004, 68-72.

bottle a way of life, a way of being, and make it portable.

The ancient Israelites would become the Jews, the People of the Book, but not a book the way we think of it. These new genres of literature were not written by one author but were distilled from the wisdom of many rabbis and from generations of women and men who lived Jewish lives. They were not meant to be read sitting in a library, silently scanning the page. You needed a teacher to guide you, and a partner with whom you could discuss them. You studied them aloud, getting them into your bones and your sinews. This was because the rabbis knew that even if we had books, the real home of the Torah was inside the person and the community.

When you got together in the *beit kenesset*, the gathering house (which translated into Greek is "synagogue"), you could feel yourself part of that same past and future that the ancient farmer would invoke when he brought up the first fruits. Wherever ten Jews gathered to pray or study, when we used our ancient holy language, it was as if the nation—past, present, and future—were there.

So, Judaism is a tradition that starts with our agrarian roots, which taught us to appreciate life as a gift and a network of relationships. It evolved into an intellectual and religious tradition that distills the rich life-giving sap from those roots. This distillation of life is flexible and strong and ties us to our bodies, community, work, play and rest. Perhaps that's why when I was a young man ready to explore the world, I went from the lush trees, beaches, green mountains and blue water of Hawaii to the study houses of Israel, and I found as much life in those texts, and in this way of life, as my soul could desire.

I had discovered that Judaism is a resource for appreciating and enjoying life to its fullest. Judaism supported the work of tending to our relationships, to ensuring that all of our relationship partners (the land, the animals, the plants, our communities) were healthy and thriving.

It is through all these relationships that we truly find home, find our soul and connect to the Soul of the World.

Today, Tomorrow and Yesterday
Systems Thinking and Ancient Wisdom

I first discovered the modern turn toward relationship thinking back in the early 1980s through reading Gregory Bateson. Something in his approach intuitively felt right. In the introduction to his book *Mind and Nature*, he wrote:

> The pattern which connects. Why do schools teach almost nothing of the pattern which connects? Is it that teachers know that they carry the kiss of death which will turn to tastelessness whatever they touch and therefore they are wisely unwilling to touch anything of real-life importance? Or is it that they carry the kiss of death because they dare not teach anything of real-life importance? What's wrong with them? What pattern connects the crab to the lobster and the orchid to the primrose and all four of them to me? And me to you? And all six of us to the amoeba in one direction and to the backward schizophrenic in another? . . . What now must be said is difficult, appears to be quite empty, and is of very great and deep importance to you and me. At this historic juncture, I believe it to be important to the survival of the whole biosphere, which you know is threatened. What is the pattern which connects all the living creatures?[14]

---

14 Gregory Bateson, *Mind and Nature: A Necessary Unity,* (New York, Bantam Books), 1980, 8-9.

The *Pearl* and the *Flame*

I knew something was grabbing me when I read Bateson. I always knew that I was smart, but I didn't seem to always be smart in the way others expected smart people to be. I didn't pull things apart very well. I wasn't the one who asked the question, tearing down someone's theory. I tended to find patterns and make connections. So, I loved Bateson and wanted to follow his thread. It eventually led me to Ben Azzai.

As we see with thinkers like Bateson, this turning toward relationship, pattern, and system in explaining the world is often felt as a major paradigm shift by those who have discovered it. There is a fascinating history of many academic fields during the 20th and 21st centuries tracing this shift away from classic, reductionist science and toward a systems view of the world.

My undergraduate major and graduate minor, anthropology, was one of the first modern fields to start this shift. Rather than try to understand other cultures by picking out items, customs, and beliefs in isolation and explaining them away, as had been done (and is still sometimes done), anthropologists started to realize that cultures can only be understood in their own contexts. One must see the relationships in the whole. In classic 20th-century anthropology, this trend was expressed in different ways by the various schools and scholars in forms such as structuralism, in culture as text, and so forth. Here is Claude Levi-Strauss, one of the greats of 20th-century anthropology, writing about myth:

> ... (W)e have to read the myth more or less as we would read an orchestral score, not stave after stave, but understanding that we should apprehend the whole. ... Something acquires meaning only if one considers that it is part and parcel of what is written below on the second stave, the third stave. ... We have to understand that each page is a totality. And it is

only by treating the myth ... as a totality, that we can extract the meaning out of the myth.[15]

More recently, many other fields have started to shift toward a non-reductionist, pattern and relationship approach to understanding. The approach goes by many names: General Systems Theory, Cybernetics, Complexity, Chaos Theory, Emergence, and more. Each designates a slightly different angle, analysis, or academic approach with its own history, heroes, and subtle differences. What has seemed most accessible and useful to me is systems thinking, or simply a systems approach. Here's an example of a practitioner of one of the hard sciences describing the shift:

> Systems biology ... requires a quite different mind-set. It is about putting together rather than taking apart, integration rather than reduction. It starts with what we have learned from the reductionist approach; and then goes further. It requires that we develop ways of thinking about integration that are as rigorous as our reductionist procedures, but different. This is a major change. It has implications beyond the purely scientific. It means changing our philosophy, in the full sense of the term ... the systems level view of life can be compared to music.[16]

Something wonderful is happening in this turning, this shift toward a systems view of the world: people are finding that it links them up with ancient traditions. The tradition that I know best is Judaism, but others have noticed similar opportunities in their traditions. Buddhist scholar and activist Joanna Macy, for example, has written about the convergence of General Systems

---

15 Claude Levi-Strauss. *Myth and Meaning*, (London, Routledge), 2001, 20.
16 Denis Noble, *The Music of Life: Biology Beyond Genes* (Oxford, Oxford University Press), 2006, xi.

The *Pearl* and the *Flame*

Theory (that's what people were calling it in the 1980s) and Buddhist thought. Her integration has fueled an activism and world-view that bridges science and religion, old and new.[17]

Writers such as Rupert Ross have explored how aboriginal attitudes toward justice and education are deeply relevant to our modern search for a more effective and humane justice system.[18] Many others could be listed. The common denominator is a way of thinking that puts things together, finding patterns and relationships.

As mentioned in the Introduction, this book focuses on three core concepts of systems thinking: emergence, nestedness, and bifurcation events, or "tipping points," which I map onto corresponding core concepts in Judaism, *minyan, mikdash, and mitzvah:* the "Three Mems."

Emergence can be defined as "the whole is greater than the sum of its parts." Some qualities emerge in the whole that cannot be found in just the parts. A forest regulates its energy intake and output as all it's parts play their interactive role in something that cannot be found in any one tree or animal or plant. A play or a piece of music comes together from disparate parts into something that feels alive and takes on a life of its own, beyond the creator's power to control. In evolution, life emerged when certain molecules started working together to regulate the flow of energy into a new being. Traditional Jewish culture has many examples of emergence. One is minyan—the idea that ten adult Jews together form the minimum of Jewish community. Especially in the context of prayer, they can say and do things that weren't allowed for individuals. The community emerges when there is a minyan.

17 See Joanna Macy, *Mutual Causality in Buddhism and General Systems Theory: The Dharma of Natural Systems* (New York, State University of New York Press), 1991.

18 Rupert Ross, *Returning to the Teachings:Exploring Aboriginal Justice* (Canada, Penguin), 2006.

Nestedness, or fractals, stand for the way nature is structured in levels of organization, separated by permeable boundaries. This is like the mitochondria within a cell, within an organ, within a body, within an ecosystem, etc. Also, a family within a community, within a city, within a country, within a region, etc. Nature and culture abound with nested patterns, such as the Fibonacci spirals in nautilus shells, the phyllotaxis spiral patterns of plant leaves, the architecture of the Greek Parthenon, and the swirl of spiral galaxies. An example of nestedness in a Jewish context is mikdash, which means sanctuary. The classic example of sanctuary in ancient biblical religion is the Temple in Jerusalem. At the center of the Temple was the Holy of Holies, which was within the Holy Palace, which was in the priests' courtyard, which was within the larger courtyard, within the walls, within the city, etc.

The third principle from systems thinking is bifurcation events or, more popularly, tipping points. This is the quality of complex systems to jump and change suddenly, not in a linear, predictable way. A small action can bring about a huge result, but one can't predict if or when that shift will happen.

From Jewish culture comes the example of mitzvah—popularly defined as a "good deed" but more accurately as a "commandment." This can be any discreet action that is done as a religious act because it is mandated. Helping an elderly person carry groceries across the street would count because that is fulfilling a mandate to do kind acts, but so would lighting candles on Friday night or fasting on Yom Kippur. Any of these small actions could make a big difference in someone's life or in the world. It is also possible that they wouldn't do much at all. One never knows, and that is part of the meaning of the act: one does it because it is the right thing to do, it's a mitzvah. It may change the world or it may not. You never know.

The three Mems represent a distillation of an ancient, organic way of thinking, as it came through the Israelite tribes, the rabbis and simple Jews throughout the centuries practicing the Torah as they lived their lives. Like much of ancient Jewish wisdom, the three Mems are not always practiced or appreciated today. But I believe that when understood, they hold a key to bringing Judaism into a new relationship of harmony with systems thinking and the leading edge of our Western culture. And that can be revolutionary.

PART TWO:
Minyan—Emergence and Life

Song Circles and Shabbat Meals

My Jewish Emergence

In the summer of 1973, a shy 15-year-old kid from Hawaii (me) arrived at a Reform Jewish summer camp, Camp Swig in Big Basin, California.[19] My parents, New York Jewish intellectuals, moved to Hawaii when I was a baby. They wanted nothing more than to get away from the "New York" part of their identities, and were pretty unconcerned with the "Jewish" part as well. To say I didn't have much Jewish background would be an understatement.

But Camp Swig hit me like a revelation. There was something magical about it. I could feel it on Friday afternoons especially, when we sang together in a big circle under the redwoods, everyone cleaned up and dressed in white. I loved the sound of the Hebrew songs. I loved the discussions and earnest arguments about God, social justice and identity. I loved the artistic creativity and do-it-yourself ethic, as when campers painted the water tower or when we dug our own "gaga" pit into the ground instead of just turning over some benches. I loved the camp friendships. The fact that we were all thrown together in an intense, highly interactive social environment, away from the cliques and in-groups of home, made it easy to talk, easy to interact (alright … easier for this shy kid to meet girls). That magical mixture of camp life changed me, as it has so many Jewish teens. It made me finally feel whole. When I graduated high school and got too old for summer camp, I found that I needed to search for whatever it was I had found at Camp Swig.

---

19 A similar version of this description of my experiences at Camp Swig, the Orthodox world and my discussion of emergence appeared as "The Magic of Emergence," in *Tikkun Magazine*, Fall, 2017, 12-16.

The *Pearl* and the *Flame*

It wasn't a straight path, but the next time I deeply felt that wholeness was in my early 20s, in the Orthodox Judaism of Jerusalem, especially on Shabbat. Shabbat in the Orthodox world wasn't the same as at Camp Swig—there was no swaying in a big circle with our arms around one another singing Hebrew songs— but it came to epitomize for me much of what attracted me to Orthodoxy. On Shabbat I felt I was entering into something that held me gently for 25 hours, transforming my experience of time and self. I especially loved the Shabbat meals. They were almost always with a group of friends, maybe with a family or two. There wasn't anything wrong with small talk, catching up and joking around, but it was also perfectly normal for someone at the table to open up a book of commentaries on the weekly Torah portion and read a little, give a bit of his or her own thoughts on it, and start a discussion.

After a good discussion, we sang. The singing of Shabbat songs (*z'mirot* in Hebrew) is true folk music in the sense that tunes are composed, learned and passed on from person to person. Sometimes we'd lose ourselves in the singing and it would go on for hours.

Put together with good food and a little rest (sometimes I'd take a short nap right in the middle of the meal), the Shabbat meals of my Orthodox days were usually restorative and often richly satisfying experiences. Deep friendships were created, personal journeys were navigated, and I came to understand why the early 20th-century Zionist writer Ahad HaAm (Asher Ginsberg, poet, philospher, 1865 - 1927) said, "It is not that the Jewish people kept Shabbat, but Shabbat has kept the Jewish people."

What did the experiences at Camp Swig and Jerusalem have in common? Both had the same magic—the magic that happens when people come together long enough, or intentionally enough, to let down their guards, to share something of themselves, to enter into a circle (literally or metaphorically) of something larger than themselves.

This feeling of Shabbat as an entity, as a palpable reality that one can enter into, emerges from the community of people practicing it. I'm reminded of a story from the Talmud (Babylonian Talmud Tractate Shabbat 119a) in which the Roman emperor, having experienced a Shabbat meal with the rabbinic community, asks Rabbi Yehoshua, "Why is it that the Shabbat meal has such a delicious fragrance?" Rabbi Yehoshua answers, "There is a particular spice that we put in it." The emperor orders Rabbi Yehoshua to give him some of the spice. But Rabbi Yehoshua has to tell him: It works if you're keeping Shabbat, but if you're not, the spice doesn't work.

I believe Rabbi Yehoshua is saying that you can't take one thing—such as the food that tastes so good on Shabbat—and "bottle" it, taking it out of the context of the whole of which it is a part. It doesn't taste the same. A meal, as we know from many writers such as Michael Pollan, Claude Levy-Strauss, and Mary Douglas, is not simply putting food into your mouth. A meal is a cultural creation, something that emerges from the people around the table, the feelings and memories attached to the foods, and the tempo and rhythm of the event.

I've found that to be especially true of the Shabbat meal. We can eat challah and have wine, we can even light candles and make a blessing, but if there isn't enough context to make it a whole Shabbat experience, those individual Shabbat items by themselves can feel flat and lifeless. Perhaps even worse, they can feel kind of nice, sort of meaningful—and then we can conclude that this is all Shabbat is: kind of nice, sort of meaningful.

Unfortunately, too much of contemporary Judaism falls somewhere into that zone between "flat and lifeless" and "kind of nice, sort of meaningful." We're like that Roman emperor, trying to reproduce something by pulling out an element or two but missing the whole that emerges when all the parts come together. The point of the Talmudic story is that there isn't one "spice" that makes Shabbat special, but there is something essential, ineffable, even mysterious that emerges from the whole.

I am no longer Orthodox and too old for summer camp, but I find myself searching for the magic of wholeness that I found in those contexts. I think that's true of others, too. For some Jewish people it may be camp-like immersive experiences such as retreats, going to a Kallah (a multi-day Jewish learning gathering) or a Shabbaton (an intensive Shabbat/weekend experience). For others some version of the weekly traditional Shabbat experience can capture a lot of the magic. This isn't the property of any one form of Judaism. I've felt some of this in every movement and in many different synagogues and homes. Beyond the Jewish context, churches and mosques have a similar dynamic; clubs and teams, family gatherings and political activism. The key is creating enough of a deep context for sharing and community that there is a palpable feeling that something new has emerged. That "something new" has a feeling of wholeness—a wholeness that is greater than the sum of its parts.

Emergence: The Modern Concept

"The whole is greater than the sum of the parts" is an old adage, but it could be one of the most important keys to a healthy and meaningful life. People yearn to be a part of something. When we are a part of something, we feel whole. When we see how things connect and relate to form a whole, they make sense and resonate. They come alive.

In Western academic and scientific settings, the new discipline of emergence is gaining acceptance in many fields including biology, entrepreneurship, sociology, medicine, activist organizing and more. Experts offer many and conflicting definitions, but the simplest way to express emergence is in the old saying, "The whole is greater than the sum of its parts."

Water is H_2O, a molecule made up of hydrogen and oxygen. Neither hydrogen nor oxygen by itself feels wet or puts out fires. But together they do. Something new has emerged when the parts become a whole. An African termite mound, sometimes rising more than 20 feet over the savanna, maintains a steady temperature through the cooperative efforts of millions of termites. No one of them knows how to do this by itself, but the quality emerges in the whole colony. Closer to home, I experience emergence in teaching all the time, though it's not as consistent as I'd like.

Emergence in Teaching

Here are two true stories:

The first story: I'm in front of a group of about 40 congregants gathered in a beautiful home for a Friday night Torah discussion. Their ages range from early 60s to 80s, and our subject is "Jewish Perspectives on Aging." I give a short introduction for maybe three minutes, and then people take turns reading a few short quotes from Jewish sources related to aging. One reads, others comment, and I add to the discussion with a clarifying comment. When I sense we've had enough of one quote, we read another, and the discussion rolls on. The whole event goes on for about 45 minutes, and when I call it to a close, people break out in spontaneous applause. I'm pleased but a bit surprised, since I didn't say very much.

Another story: I'm sitting around a table with a group of congregants, teaching about the Passover Seder. I think I have some pretty interesting things to say about the subject, but as I launch into my lesson, I notice something is wrong. I am droning on and people are drifting off. I can practically hear the energy leaking out of the room, like air from a punctured tire. My wife, who happened to be visiting that class, looks on in horror, trying to figure out a way to get my attention and stop the bleeding. Fortunately, someone asks a question, cutting short my monologue. Perhaps I have a second chance.

In the first story, I created a container for something to emerge. It was a "chemical" reaction between the quotes, the people in the room, and me, the teacher, holding the boundaries and setting the stage. I stepped back enough to let a process happen between the people reading and reacting to the texts and to one another. Ideas I hadn't thought of came out of the mix, and people felt creative and involved.

In the second story, I created a linear, one-directional pipeline from my brain to theirs. Maybe in my mind the material was interesting, but the process was static. I had forgotten the most basic lesson of teaching: It's not just the subject, and it's not just the teacher; something has to happen in the space between the leader, the people in the class, and the subject matter.

Like so many things in our world, a class or a facilitated discussion is an emergent phenomenon. It does not exist in any one thing or person, but rather in between, in the interactions. You can't see emergence, but you know when it's there and when it's not.

Emergence of a Paradigm Shift

It is important to recognize that emergence as a modern, scientific concept is bucking the trend of more than 300 years of scientific thought. Since the days of Descartes, Bacon, Newton, and other pioneers of the 17th-century Scientific Revolution, we have been taught that the scientific way to understand the world is to break things down into component parts. Emergence is one concept within the growing area of systems thinking, which seeks explanation by putting together rather than breaking apart. While there have been a number of discrete schools of thought and different labels for the movements that have parted from reductionism, these have been like waves lapping up on a shore even as the tide is going out.

I believe that we are coming to a tipping point (another concept spun off from systems thinking). The tide is turning. We are coming to a point where the old reductionist explanations are not adequate to the problems that we face, and thinkers in many different fields are turning to a new systems approach. This approach is much more in sync with traditional cultures such as Judaism.

One clear example: It has been very powerful to know that diseases are caused by particular bacteria that we can see and identify with a microscope. That we can now make antibiotics to fight those bacteria has been a revolution in human health. But, for all the success modern medicine has had through breaking things down and looking for the most basic underlying causes of disease, we are starting to see the limitations of this reductionist approach. We've started to see that killing the bacteria isn't a one-way street, it's a feedback loop: the more we use antibiotics, the more the bacteria mutate to resist them. We are now in real danger of "superbugs," antibiotic-resistant strains.

We have also learned that, while it is true that particular bacteria can be a major factor in the onset of disease, many other factors also play a part. For one thing, bacteria are present all the time, but we aren't always sick. We now know other bacteria might be counteracting the "bad" ones that are living in our bodies. In fact, we have discovered that the human biome (the community of microorganisms living inside us and on us all the time) is essential to our health and functioning.

We also know that our immune system fights off many bacteria. What factors, such as emotional stress, might be weakening our immune system? What factors, such as strong social bonds, might be strengthening our immune system? We begin to see that health is a complex system. It emerges from the interaction of many factors—biological, psychological, social and spiritual—that together create a whole greater than the sum of its parts.

Minyan—Emerging into Community

The ancient Jewish idea of minyan can stand in for and point us toward the myriad ways that emergence runs through traditional Jewish life. A minyan is a quorum of ten Jewish people over 13 years old (for most of Jewish history only men were counted; now, except in Orthodox communities, all genders are counted). Shabbat is not the only example of emergence in Jewish life. I've found emergence in Jewish texts in the way they combine and come together to create meaning, and even emergence in God. God's essence may be unknowable and completely beyond our grasp, but the way that God manifests in our world, especially our experience of God, has a lot to do with that mystery of the "something more" we now call emergence. The common denominator is that we feel something come alive when all the ingredients come together.

More than the Sum of its Parts—
Holiness and Wholeness

Minyan is encoded in Jewish law and practice as a new entity, more than the sum of the individuals, that emerges when there are ten Jews present. The most common way we experience minyan is in prayer community. We need to have a minyan in order to say Kaddish, the ancient Aramaic call-and-response prayer recited by mourners. A minyan is necessary to say the call-and-response Barchu blessing which gathers the community to pray. We say this same call and response when we read from the Torah scroll.

The traditional term for these parts of the prayer service is *d'varim she'b'kedushah*, holy words. So, holiness and community are related. Community creates a new kind of wholeness, a vessel for holiness, that goes beyond the individual. Something qualitatively different occurs when a community comes together; we come into our humanness more fully through communal expression. Wendell Berry wrote beautifully about wholeness and holiness, and about our communities, even beyond the human:

> … (T)he concept of health is related to the concept of wholeness. To be healthy is to be whole. The word health belongs to a family of words, a listing of which suggests just how far the consideration of health must carry us: heal, whole, wholesome, hale, hallow, holy. And so it is possible to give a definition of health that is positive and far more elaborate than that given to us by most medical doctors and the officers of public health.
>
> If the body is healthy then it is whole. But how can it be whole and yet be dependent, as it obviously is, on other bodies and upon the earth, upon all the rest of Creation, in fact? . . . Our bodies are also not distinct from the bodies of other people, on which they depend in a complexity of ways from biological to spiritual. They are not distinct from the bodies of plants and animals, with which we are involved in the cycles of feeding and in the intricate companionships of ecological systems and of the spirit. They are not distinct from the earth, the sun and moon and the other heavenly bodies.[20]

Holiness, then, is related to wholeness, but crucially, that doesn't preclude inter-dependence. Wholeness can be complex, and nested. Paradoxically, we find our own personal wholeness within our inter-dependence with others, both human and beyond human.

---

20 Wendell Berry, "The Body and the Earth" in Norman Wirzba, ed., *The Art of the Commonplace: The Agrarian Essays of Wendell Berry* (Berkeley, Counterpoint Press), 2002, 98-99.

"Am Israel Chai"—The Jewish People Lives

A minyan creates a community, specifically a community of Jews. If we dig deeper in Jewish consciousness about minyan, we find this group of ten or more Jews is a microcosm of a much larger living entity: the Jewish People, which stretches across physical borders and across the centuries.

Other traditional cultures have their own ways of imagining the idea of their peoplehood. I was deeply influenced years ago by the Native American classic *Black Elk Speaks*, in which John G. Neihardt writes, quoting Black Elk, about "the hoop of his nation"—how that hoop might be broken amid the disastrous war with the whites but would ideally be blossoming and alive.[21]

A similar feeling surrounds the idea of the Jewish People. This is especially true in the Jewish mystical tradition, where the term *"K'nesset Israel"* the gathering of Israel, is not only the Jewish People but also overlaps with an entity that is identified with the immanent, feminine aspect of God.

In the erotic symbolism of Kabbalah, the masculine, transcendent side of God, and the feminine, immanent aspect yearn to come together. When they do, the divine flow of energy brings new, vibrant life and blessings to the world. Kabbalistically, we understand the minyan of ten Jews as a "vessel" for God's presence.

God, the Archetypal Feminine and Community

The fact that *K'nesset Israel* means both the gathering of Israel and is also a Kabbalistic term for the indwelling, feminine aspect of God is confusing, but it reflects an essential idea. God emerges in communities in the same, miraculous way life emerges from what could have been a gathering of cells: the wholeness that

---

21 *Black Elk Speaks: Being the Life Story of a Holy Man of the Oglala Sioux,* as told through John G. Neihardt (Flaming Rainbow), (Lincoln and London, University of Nebraska Press, 1932, 1988.

is greater than the sum of its parts births something new. The gathering of Israel is the vessel that allows God to manifest in the world in this particular way.

The imagery of the vessel has in many cultures been associated with feminine symbols of the womb and thus with the primal Source of Life. In that regard it makes perfect sense that the collective entity, the Community of Israel is the feminine *K'nesset Israel,* which is also associated with the feminine Divine Presence, *Shekhinah.* Yet, as we know, for most of Jewish history, the minyan which represented the presence of that community was exclusively male. It seems that going all the way back to biblical times there had been a patriarchal displacement of women by men. This was more complicated than simply misogyny, a hatred of women, and a desire for power by men. Rather there was sometimes a kind of co-opting of the feminine imagery onto male roles and male institutions. There seems to have been a kind of ambivalence in the male biblical writers. They both envied what they witnessed as the feminine power of bringing life into the world—the most primal of miracles and thus one of the earliest and most powerful inspirations for religious feeling—and they also feared and tried to repress that feminine power of life.

Modern scholars, rabbis, ritualists and writers have been doing the work of uncovering important ritual roles that women did play in biblical times which were partially covered or censored by the male writers, thus helping to recover and renew the original power of the feminine and of women. Today, in most Jewish settings, all genders are counted in a minyan and we are starting to more explicitly recognize and honor feminine archetypes as they are expressed through women and people of all genders.[22]

---

22 See, for example, Jill Hammer, *The Hebrew Priestess: Ancient and New Visions of Women's Spiritual Leadership,* (Teaneck, NJ, Ben Yehuda Press), 2015, Carol Meyers, *Ancient Israelite Women in Context,* (Oxford, Oxford University Press), 1988, Elyse Goldstein, *Revisions: Seeing Torah Through a Feminist Lens,* (Woodstock, VT, Jewish Lights Publishing), 1998, Catherine Keller, *From a Broken Web: Separation, Sexism and Self,* (Boston, Beacon Press), 1986, and many, many more.

Minyan and God—Emergence is Real

One of the big questions theologians and scientists have about emergence is "Is it real?"[23] Especially in the mode of thought we've been raised in—reductionist thought that would have us explain the world by reducing things to their underlying cause—we find it hard to say that the uncanny way that the whole is greater than the sum of the parts is "real."

Rabbi Kalonymous Kalman Shapiro, The Piaseczner Rebbe, (1889-1943) was leader of the Hasidic community in the Warsaw Ghetto during the years of the Holocaust. He remained in the ghetto and continued to offer his leadership during the excruciating years of the war. He addressed this issue when writing about prayer in community, a minyan:

> This … relates to the matter of prayer in community (minyan). At first glance, the noun "community" is just a word – a gathering of people such that the individual people are present. And we give it the name "community." If we were to not give it the name "community," the individual people would still be there, but not the "community." But, in fact, … the community (minyan) is a separate essence, and it exists in heaven. This is the concept of *K'nesset Israel* as is known. When the Jews gather together to worship God, then God brings down for them the quality of "community"/*Kenesset Israel* according to their portion. Therefore, even though each Jew is holy and is obligated to serve God, still, they aren't permitted to say all the prayers as we do in community such as Kedusha, Barchu, and so forth. This is because the community pulls in the holiness of minyan, which

---

23 See, for example, *The Re-Emergence of Emergence: The Emergentist Hypothesis from Science to Religion,* Philip Clayton and Paul Davies, eds., (Oxford, Oxford University Press) 2006.

goes beyond the holiness of the individual, so they can add those prayers. So, also, when you speak words of Torah to a group, you need to have the intention to pull in the soul of the group, beyond arousing the souls of the individuals.[24]

In Shapiro's language, minyan is real, as what he calls a "spiritual entity." In all his writings and in his life—which involved unbelievable bravery and faith while leading his community in the Warsaw Ghetto—he never wavered in centering every aspect of his life around the reality of such "spiritual" things as minyan, holiness, and, of course, God. He knew very well that we humans have the choice to ignore such realities. But he knew that if we give ourselves the spiritual training to see them, they are the most real, and the most consequential realities that exist.

Our Consciousness of Minyan

Is this spiritual reality, this kabbalistic union of the transcendent God and the indwelling, feminine Shekhina also known as *K'nesset Israel*, felt by all Jews who come together to make a minyan? Clearly not, but even for the non-kabbalist, average Jew, being part of a minyan can carry a deep sense of obligation and satisfaction. When I was a part-time rabbi for a small community in mid-coast Maine, that sense of obligation and satisfaction was palpable, especially when people got a call that a mourner needed a minyan to say kaddish. It aroused a deep sense of sacred obligation that people felt as an essential expression of their Jewishness. I also saw that even that vestigial, nostalgic sense of the holiness, obligation and satisfaction of "making a minyan" was fragile and liable to be lost in the younger generation.

Up in that small town in Maine, it wasn't always easy to get a minyan. On an average Friday night when there wasn't a special

---

24 Kalonymus Kalman Shapiro, *Sefer Derekh HaMelekh*, (Jerusalem, Va'ad Hasidei Piaseczna), 1995,"Parshat Hayei Sara," p. 13-14. The translation is mine.

need to support a mourner, it wasn't unusual to wait a bit nervously to see if we'd get a minyan. And there was always the pressure to simply abandon or modify the rule of ten.

Our experience told a story of fading consciousness of that sense of obligation, of peoplehood. It's one of the symptoms of living in our individualistic Western culture. That consciousness has been fading faster of late, but the process is several centuries in the making.

Minyan: Eastern and Western Style

A friend of mine once said to me that he thought the difference between how Ashkenazi Jews, those with origins in Europe, and Sephardi or Mizrachi, those from the Spanish diaspora of North Africa or the Middle East, experience minyan. He said, "An Ashkenazi is asking, 'What am I getting out of this minyan?' where the Sephardi Jew is simply enjoying being together." When I first visited Israel as a student, I lived in a small town and got a taste of this.

Most residents of this town were Moroccan Jews who had emigrated to Israel and were "settled" (really, more like abandoned) in places like this in the middle of the Negev desert. By the time I arrived more than a decade later, they were still adjusting and trying to make it in this new society. A soft drink bottling factory was the basic employment, and while most of the younger generation was secular, the small town held probably 15 Moroccan synagogues.

I got some of my first experiences of helping to make a minyan with the mostly older Moroccan men who came to synagogue early in the mornings. I'd borrowed a pair of tefillin, the leather boxes and straps that are traditional for men to wear during morning prayer, and would put them on my arm and my head every morning. While I did that, the men would be reciting Zohar, the central text of Kabbalah, not for the deep, mystical poetry that I'd discover years later, but as a devotional ritual as they got ready for prayer.

The physical plan of the synagogue was very different from the Western ones I occasionally had attended back in the U.S. Their synagogues reflected a traditional pattern of people all facing one another, essentially a big circle, facing both inward and outward. The men lounged on thinly cushioned benches lined up along the walls and faced each other from the middle around the bima, the raised platform in the center of the room. The bima was where the service was led, although for certain prayers, the person leading might be anywhere in the room. On a Friday night the different verses of L'Cha Dodi might be sung in a beautiful, lilting Middle Eastern chant by any of the men, or even boys, around the room. I also found that in this community the circle of minyan, this emergent circle of belonging, didn't stop in the synagogue but characterized their whole way of life.

I had been assigned a "host family" in my capacity as a volunteer. This particular family had been assigned to me because their older son, Avner, played in the Youth Club basketball practices that I coached.

They were fairly typical of the local Moroccan families. The father worked in the bottling factory. The mom also worked, I think, doing cosmetics, but also sometimes in the factory. They were "secular" or perhaps "traditional" on the Israeli social-religious map, since only strictly Orthodox folks would be called "religious"—but they were pretty religious to my eyes.

Every Friday night they invited me to eat the traditional Shabbat dinner with them at their apartment. Well, that's not exactly true. After I ate with them a couple of times, they would just expect me to show up for Friday night dinner. I was now part of the family, why would I need an invitation? One week I didn't go and was surprised to see their youngest son appear at my apartment with a plate of traditional spicy Moroccan fish, their sweet, dense, Moroccan challah bread, and the other foods we always ate. The concerned family had assumed that if I didn't show up, I must have been not feeling well.

They were, as much as their synagogues, living in a minyan kind of way, with community and family as the foundation of everything. That was just the way it was, and there was a satisfaction and pleasure in that sense of belonging and order.

Of course, not all of this way of life easily translates into our Western, more egalitarian life-styles. Their family was quite patriarchal, with what we would recognize as traditional gender roles. While in the modern Western context most women prefer a wider choice of roles, in this fairly traditional, Moroccan culture (although they may also have chafed at the patriarchal restrictions) the women I met seemed to take pride and satisfaction in being what anthropologist Susan Starr Sered describes as ritual experts. They were dealing with the embodied areas of life outside the male dominated synagogues. Sered notes that Shabbat and holiday food preparation, rituals of birth and death, charity and more were critical parts of the religious life and the elderly, Mizrachi women in Jerusalem that she interviewed took great pride in their religious roles.[25]

---

25 Suan Starr Sered, *Women as Ritual Experts: The Religious Lives of Elderly Jewish Women in Jerusalem,* (Oxford, Oxford University Press), 1992.

Life Emerges

Feeling part of a minyan, we feel the life of the community. The minyan points us to something whole: an entity with an inside and an outside, such as a circle. This property of having a boundary between inside and outside may define the difference between life and nonlife. Molecular physicist Harold Morowitz postulates that the origin of life was precisely this creation of a boundary: "It is the closure of [a primitive] membrane into a 'vesicle' that represents a discrete transition from nonlife to life."[26] That membrane or vesicle, he explains, creates the conditions for a feedback loop of taking in nutrients and letting out waste in the right proportions—the beginnings of what we call metabolism—and so the world had a new level of emergence...life! Imagine that moment billions of years ago when Earth transformed from a rock in space to the home of life.

We all feel amazement when we see a newborn baby, an intrinsic awe and pleasure in seeing new life, radiating vibrant energy, animated with Something that can't be defined. We never get used to it. And when we see death, we can never completely grasp it. How can the person who not long ago was filled with stories and song and intelligence and wit and memory now be an inert corpse, a collection of empty cells soon to decay? It doesn't add up. Something has been lost, that is the essence, but we can't define that Something.

Many areas of human life such as literature and music reflect the same dynamic as biology. Some kind of context encloses and holds the individual elements together enough for them to interact in a feedback loop such that when one element changes, there is

---

26 Morowitz, Harold. *Beginnings of Cellular Life*. Yale University Press, 1992. Pg. 9. Quoted in Fritjof Capra, *The Hidden Connections*. Anchor Books, 2002, 22.

a reaction from another element, and that reaction comes back to the first, affecting its next move, and so on. Imagine a play that has been rehearsed and the actors are saying their lines perfectly well, but there's something missing that is hard to put your finger on. Maybe the director calls actors together and encourages them to open their hearts, listen more deeply to each other, then run the scene again. This time something clicks. A new energy has mysteriously entered the room and everyone can feel that this time the play has come alive.

I recently had the experience of working with a master musician and teacher, Joey Weisenberg, who created the conditions for music to come alive. A couple hundred people gathered in a room. We first learned a melody; he led us through the notes, then the rhythm. Little by little it sank into our heads, bodies and mouths until we could stop thinking about it and just sing. Then suddenly we were singing together, a room filled with song that we together created. It came alive.

adrienne maree brown writes in her book *Emergent Strategy* about how activist groups, committees or organizations can imitate the emergent quality of a flock of birds or a school of fish moving together. "Murmurization" is the word for that magical shifting, turning, instantaneously adjusting movement of hundreds of individual beings moving together. Each creature needs to feel the subtle changes of the one next to it up, down and to the side, reacting to each tiny shift. So, she says, a group needs to trust, listen and open to one another so that their committee or organization comes alive. [27]

Not all elements of the emergent feedback loop need to be real. They can be fictional. Novelists will tell you that they get to a point where it feels as if their fictional characters are leading them. A writer creates the characters, but they take on a life of their own. That is when the author can feel the novel coming

---

[27] adrienne maree brown, *Emergent Strategies: Shaping Change, Changing Worlds* (Chico, CA, AK Press), 2017, 71.

alive. I feel the same thing when I draw or paint. I love the feeling of looking out into the world and then putting some ink down on paper, then looking again, making another line, and feeling the interactive dance between my eyes, the scene, my pen and the paper, all working together. Any artist has known the paradox of creation: Your best work happens when you feel that it leads you as much as you lead it. That's the emergence of a living work of art.

A Biblical Example— Tearing the Fabric of Society

When Moses in Deuteronomy exhorts the people to "choose life," he's capturing a current that runs strongly through the biblical worldview and gets carried down through the rabbis and into the Jewish ethos, up to the classic toast, "L'Chayyim! (To Life!)" This concern for life is not simply with the most literal level of preserving physical life, but it points to all the other levels of life that we've been seeing are born through the magic of emergence. They knew that life isn't simply a heart beating or breathing in and out. We can be more or less alive, and a person or a society can have death creep up on them before the physical demise of their bodies. For example, the rabbinic phrase to denote the sin of embarrassing someone in public is *halbanat panim* – whitening the face. The Talmud understands that when someone is embarrassed, the blood runs out of their face and they look as if they had died. The Talmudic sages even go so far as to state, "whoever whitens the face (i.e., embarrasses) his friend in public, it is as if he had shed blood." (Babylonian Talmud, Bava Metzia 58b)

Shaming and its association with death was well known to both the rabbinic and biblical writers. Especially in small, face-to-face societies, shaming, libel, and malicious gossip can not only hurt an individual, but can weaken the group as a whole. It can tear the social fabric.

63

The *Pearl* and the *Flame*

Social anthropologists such as Mary Douglas, especially in her 1970 book *Natural Symbols: Explorations in Cosmology*, have pointed out how the condition of the social structure may be symbolically represented on our bodies and, by extension, our clothing and even our homes. For example, a highly controlled and hierarchical social structure such as the military is reflected in the short haircuts, beardless faces and drab, identical uniforms of soldiers. The opposite of this might be an artists' commune, where long hair, beards and colorful, loose and varied clothing is the norm. This anthropological insight opens our eyes to some broad and central themes in the Torah.

Wild Hair, Torn Clothing and Death

Many Jews today are aware of the ancient mourning practice of tearing one's clothing upon hearing of the death of a close relative or beloved teacher. A pale shadow of this remains in the modern funeral practice of giving mourners a black ribbon to tear at the funeral. Literally tearing one's shirt still happens in Orthodox and a few other observant communities.

Traditionally observant Jews also don't get a haircut for the first 30 days after the death of a relative and during certain periods of the year that are considered times of collective mourning. These practices go back to biblical texts that connect tearing one's clothing and letting one's hair grow "wild" with death. But the first time we hear about this practice in the Torah, it's to give an example of the opposite: Kohanim, the priests, were prohibited from doing either of those things. After the tragic death of the sons of the High Priest Aaron, Moses warns Aaron and his other sons:

> Do not bare your heads and do not rend your clothes, lest you die and anger strike the whole community. But your kinsmen, all the house of Israel, shall bewail

the burning that the LORD has wrought. (Leviticus 10:6, JPS)

The word that the JPS translation renders as "bare" could also be translated as "make wild/disorderly." Death is the opposite of wholeness. Death is the ultimate brokenness, and that brokenness is emotionally and physically expressed on our bodies and clothing. We tear our clothing and let our hair grow wild to show that we are broken. Death has touched us as well.

But the Kohanim are not regular people. They hold a position in Israelite society that makes their bodies and their clothing not only their private matter. In ancient Israel, priests' bodies stand for the wholeness and life of the Israelite society. Thus, "Do not bare your heads and do not rend your clothes, lest you die and anger strike the whole community."

The High Priest had even more strict rules about body and clothing. He wore a breastplate that held the 12 stones, each representing one of the 12 tribes. He could not take on the mourning rituals even for his closest relatives. In some ways we can think of a CEO who needs to wear a business suit and have neat hair because the CEO represents the company, while the "creative team" can come to work in jeans and T-shirts (except, of course, in the case of a tech company where that kind of order and hierarchy is explicitly rejected, and the CEO does represent the ethos of the company by wearing jeans and a T-shirt).

Once we start to pull on this strand, we can see that this symbolism of wholeness/brokenness on the body and clothing is woven (braided, perhaps) throughout the Torah. The opposite of the Kohen, Priest, is most certainly the person with *tzara'at* which has been translated as leprosy, but which scholars agree wasn't actual leprosy. We don't know exactly what it was. My teacher, Jacob Milgrom, and others after him have called it "scale disease" because its main symptom was redness and scales on the body. Not only are these scales a brokenness in or on the body, but the *metzora* (the noun for one with *tzara'at*) was instructed to separate from the rest of the encampment, let their hair grow wild and tear their clothing.

Later rabbis picked up hints in the biblical text that the *metzora* brought this upon themselves by speaking slander or spreading gossip. The most famous case of this was Miriam contracting the disease after she and Aaron spoke against Moses (why Aaron didn't get it is an interesting question, perhaps related to some of the gender matters we're about to discuss). The Talmud says of the *metzora*, "He separated husband from wife, friend from friend, so now let him sit separated from the community." [Babylonian Talmud, Arakhin 16b]

The mention of separating "husband from wife..." isn't random. Another strand in this pattern is the strange and disturbing case of the *sotah*, or suspected adulteress. Parallel to the Kohen, whose body stands as a symbol for the wholeness and integrity of the Israelite nation, the married woman's body played the same role in the family. Just as the body and clothing of the priest, especially the High Priest, needed to be whole and intact in order to embody the life of the nation, the bodies and clothing of women stood for the wholeness and life of the family unit. As discussed earlier, it seems probable that women were the original source of this reverence for the life-containing-vessels, and that the male priests were claiming that feminine symbolism for their own.

In the case of the *sotah* ordeal, not only were women denied the high social status given the male priests as symbols of wholeness and life, but they were put into a terrifyingly vulnerable position as they bore the brunt of blame for any perceived break in that wholeness. Women were subjected to the shaming *sotah* ordeal if her husband for whatever reason became jealous.

A suspected adulteress was brought before the priests for the *sotah* ordeal. With no clear evidence or witnesses, there couldn't be a trial, yet the trust and wholeness of the family had been disrupted—mainly by the husband's jealousy. So, this ritual was done to bring things back to equilibrium. In the ritual, the name of God is written on a parchment, torn up, mixed with the dust on the ground, put in water, and given to her to drink as a testing potion. As a part of the ordeal, her hair is made wild or uncovered (*parua* the same Hebrew word used in the case of the priests mentioned above) and her clothes are torn. The break in the trust and unity of the marriage shows up ritually on the woman's body and clothing.

As Rabbi Kohenet Jill Hammer, Kohenet Taya Shere and others have shown, it is possible to recover the priestly roles that women did play in earlier periods, and which have so often been co-opted and censored. The connection between life and wholeness in the family, community or people which we find in these biblical texts does not need to be expressed today in their ancient patriarchal forms, but can be renewed and given new contemporary expression.[28]

Sometimes the family metaphor was applied to the national level as we see in a kind of "national *sotah*" in the incident of the Golden Calf. When Moses comes down from Mount Sinai and sees the people dancing around the idol, he says that they are "wild" (*parua*). Then he takes the molten calf, grinds it up, mixes it with water, and makes the people drink it as a potion. The people at Mount Sinai were supposed to have been entering into a kind of marriage with God, but instead were unfaithful with the calf.

More strands could be added to this pattern of wholeness and brokenness that runs through the biblical and rabbinic texts. The family is the basic unit of fertility, and the life of children is not only a biological fact but is tied to the sense of wholeness and trust that makes the family a source of life. When the High Priest goes into the Holy of Holies on Yom Kippur, he gets a blessing of fertility for the nation's families, but also for animals and crops. Rabbi and scholar Bonna Haberman wrote on ways of recovering that feminine sense of the Holy of Holies as a feminine enclosure in our present day Yom Kippur afternoon ritual recounting the

---

28 See Hammer, *The Hebrew Priestess,* for many examples.

High Priest entering the Holy of Holies.[29]

The people as a whole need to be a fitting container for the blessings of life. The ancient Israelites lived in a world in which all parts of life were related. Their minds were much more attuned than ours are to the metaphors of connection that tied together agriculture, family, plants, animals and humans into one living world. But we, too, can learn to make these connections.

29 Bonna Devorah Haberman, "The Yom Kippur Avoda within the Female Enclosure," in *Beginning Anew: A Woman's Companion to the High Holidays,* ed. Gail Twersky Reimer and Judith A. Kates, (New York, Touchstone Books), 1997, 243-57.

Agriculture, Culture,
Health and Holiness

The juxtaposition of elements that react to one another in a feedback loop is a necessary condition for life and for the kind of "aliveness" that we've been discussing. One of my early heroes of this kind of thinking, Gregory Bateson, argued that all living systems have "mind" in as much as they all involve actions and reactions within feedback loops of information. A context, a membrane or even a literary juxtaposition that creates the possibility for feedback between elements: this brings things alive, and it is essential for the emergence of life. Life always exists in patterns.

As mentioned earlier, one of Bateson's most famous phrases was "the pattern which connects." Bateson crossed disciplinary boundaries, making ground-breaking contributions to anthropology, biology, psychiatry, education, family therapy and more. His thought wasn't captured by any one of those disciplines because he recognized that all life, whether human, plant, animal or ecosystem, runs on the patterns which connect.

Patterns are an indication of complex interaction, feedback, life, and health. So, it is no surprise that another one of my favorite writers on nature and culture, Wendell Berry, also makes pattern a central theme in his writing. One of Berry's classic and often quoted essays is "Solving for Pattern." In this passage from "Solving for Pattern," we get an illustration of the importance of pattern and how it is related to health. According to Berry, health emerges in the same way that a pleasing pattern emerges from the balanced, complex interaction between parts.

The real problem of food production occurs within a complex, mutually influential relationship of soil, plants, animals, and people. A real solution to that problem will therefore be ecologically, agriculturally, and culturally healthful.

Perhaps it is not until health is set down as the aim that we come in sight of the third kind of solution: that which causes a ramifying series of solutions – as when meat animals are fed on the farm where the feed is raised, and where the feed is raised to be fed to the animals that are on the farm. Even so rudimentary a description implies a concern for pattern, for quality, which necessarily complicates the concern for production. The farmer has put plants and animals into a relationship of mutual dependence, and must perforce be concerned for balance or symmetry, a reciprocating connection in the pattern of the farm that is biological, not industrial, and that involves solutions to problems of fertility, soil husbandry, economics, sanitation – the whole complex of problems whose proper solutions add up to health: the health of the soil, of plants and animals, of farm and farmer, of farm family and farm community, all involved in the same interested, interlocking pattern—or pattern of patterns.[30]

Berry describes some of the basics of emergence. There is a limited context, the small family farm, where parts can be put into complex interdependent relations with one another to form a whole. As he points out with the example of animals and plants interacting, striving for balance and symmetry is key to the process.

---

30 Wendell Berry, "Solving for Pattern," *The Gift of Good Land* (New York, North Point Press), 1981, 136-37,

Anyone who has done this kind of balancing knows that a myriad of factors constantly shift and these affect the entire system. There is a bit more rain today, this cow is not feeling well, the wind blew down that corner of the fence. Once you enter into it, you see a dynamic, complex relationship that constantly feeds back on itself and therefore needs constant adjustment and care. This is the beauty and intelligence of the craft of farming—which keeps its practitioners dedicated and proud—as creative and alive as any work of art.

I'm not a farmer, but I've been backyard gardening and composting for at least 25 years. So, I have an inkling of what Berry is talking about on my own small scale. My wife and I are constantly tinkering, pondering and adjusting every aspect of our little operation. Rebuilding the kids' sandbox into garden beds when they grew out of it; figuring out which combination of plants can best take advantage of the sunlight, discourage the insects we don't want, and encourage pollinators we do want; pruning and spraying Ne'em oil to keep the tomatoes healthy. The list goes on and on. In general, these tasks pose interesting challenges and give us a lot of pleasure. Sometimes the problems of the garden can become almost obsessions, but in a good way—absorbing my attention and focusing my mind on some tricky problem that brings out my latent resourcefulness and my humility.

About a year ago, for example, one of the local developers decided to tear down a large part of the town's commercial area about a mile from our house and rebuild it bigger and better. But the closing of a few restaurants and the disruption of all that construction caused swarms of rats to abandon their normal territory and go looking for more peaceful and plentiful hunting grounds. After having composted in open chicken wire bins for years without any major rat problems all of a sudden, our neighbor on the other side of our backyard fence was complaining about rats coming from our compost. Someone (not sure which neighbor that was) even called the city health department. The health official came out and said I'd need to fix the rat problem or stop composting.

Since composting is literally a religious obligation for me, an almost daily practice that grounds me and keeps me connected to the earth, not to mention providing rich soil every year for our garden, that warning sent me into action. I felt a bit like the hapless Wily Coyote from the old Road Runner cartoon, as for weeks on end the rats outsmarted me and undid my MacGyver-esque inventions meant to keep them out of the compost.

New, hard plastic bins were chewed through. Putting the bins on top of thick wooden planks didn't work. Wiring the bins down to the planks didn't keep the rats out either. Neither were the rats much discouraged when I put sharp staples sticking out all around the bin's edge. I enlisted several neighbors' help, and found myself balancing precariously twenty feet up while putting up a couple of owl houses in nearby trees to attract the rats' natural predators. No owls have come so far (we theorize that too many people have been putting out rat poison and, of course, the rats are doing fine but the owl population has declined).

We finally realized that the rats weren't living on vegetable scraps alone and they must be getting into the garbage to get protein like cheese and meat. The compost bins were only a supplement. When we put bricks on the garbage bins, we started to see a decline in rats. And then the restaurants started opening again and we're here with our newly reinforced composting operations and no rat problem.

The social and physical environment in our neighborhood, from the business district a mile away to the trees around our yard, contribute to the complex patterns that I need to consider when I'm trying to keep my compost operation turning out rich, dark soil for our garden. My wife, Ilana points out that when I'm absorbed in one of these practical, creative challenges I'm always whistling. I'm not really conscious of it, but the whistling is a sign that I'm in the flow. And that's a good sign, both for the garden and for me.

The complex interaction of parts in a feedback loop creates a pattern that is both challenging and beautiful. There is nothing nostalgic or quaint about the well-run farms that Berry describes or backyard gardens and compost bins such as ours. In fact, as Berry is arguing, these kinds of pattern solutions solve our food production problems in a much more healthy and productive way than the "normal" agribusiness solution of huge mono-crop farms.

We have bought into the lie that only on this industrial model can we feed the hungry of the world. But how efficient are we being when we are creating so many side effects as the result of producing this food? Is this really the only way to do it? Small farms produce, after all, a much larger yield per acre than industrial farms.[31] The problem is that big corporations can't make huge profits from those small farms.

I want to juxtapose that beauty and craft that Berry describes for the well-run farm with something that may seem completely different: Jewish learning. In Berry's terms, I'm suggesting a new pattern or pattern of patterns. It comes from the Piaseczner Rebbe, Rabbi Kalonymous Kalman Shapiro. Rabbi Shapira's comments focus, like Berry's, on wholeness. He writes:

> The mitzvot (individual commandments) reveal only the "limbs," while the *sefer* (book) reveals the essence, the whole structure of which the mitzvot are only the limbs. . . . .
>
> Therefore, when one looks into a book or hears words of Torah, if one only sees or hears one or two things, and especially if someone only wants to hear a nice idea or "droosh" – one only hears the "limbs"

---

31 "The difference in productivity between small farms and industrial farms is not slight. In every country for which data is available, smaller farms are shown to be 200 to 1,000 percent more productive per acre unit area." Ellen F. Davis, *Scripture, Culture and Agriculture.* (Cambridge, Cambridge University Press) 2009, 104. See her footnote 15 for her sources.

and misses the teacher's wholeness, and doesn't encounter the prophecy within the words. Therefore, this person only knows what they heard – because they didn't encounter the whole teacher, the conduit of prophecy, and they'll remain in all their actions in the same state of doubt and ignorance as before they heard anything.

Therefore, when you study, take only one or two books, according to the time you can allot to them, and study them deeply. And when you hear Torah from a teacher, return often and learn a lot from them. Because it is in the combining of all the limbs that the essence or the person's wholeness is revealed, and it is through that wholeness that God reveals prophecies. And then a person knows not only the words that one heard, but will also come to reveal *from within oneself* new thoughts and paths and understandings.[32]

His analogy between a person and a book is deeply woven into Jewish consciousness. We treat a sacred book with the respect due a person. We lovingly "dress" a Torah; when a sacred book is worn out, it is buried like a person; famous rabbis in Jewish history are often known simply by the name of their books; we sometimes refer to a learned person as a walking book. Neither a person nor a book can be reduced to the sum of its parts.

Rabbi Shapiro puts us into relationship with that wholeness: When you take in that essence, you can then reveal from within yourself new insights. There is a transformative moment when

---

32 From *Derekh HaMelekh* by Kalonymus Kalman Shapira, Parshat Shemot. To clarify his use of the word "prophecy," he explains that although the Talmud states that prophecy has disappeared from the world after the biblical period, that only refers to the kind of prophecy that predicts future events; the "inner prophecy" (my term) of intuition into the deeper meanings of a text—for example, those moments when one feels she is reading God's thoughts—may still be accessible today. The translation is mine.

your own wholeness meets the wholeness of the book or the person from whom you are learning. What had been a linear acquisition of external facts becomes inspiration. I come into contact with a living essence and it touches my living essence. Then I become not only a receiver of the Torah, but a living wellspring of Torah.

In another part of this teaching, Shapiro writes about how this kind of insight can come upon a person suddenly:

> . . . sometimes, when a person has been studying a book deeply, suddenly they are amazed by something and it almost knocks them over, and this thing can pierce their heart and continue to affect them for years until they are changed into a completely new person, sanctifying them and raising them up.

The process of real learning isn't a linear collection of facts. It involves unexpected jumps when one "gets" that idea beyond the words, when it all comes together. The person learning is as much a part of the process as the teacher or the book one is learning from.

You need to know the whole before you can riff on the parts. It's that feeling of making it "your own," yet you're also in touch with something beyond yourself. This kind of learning connects heaven and earth when you make that quantum leap from the words on the page to a deeper essence that feels somehow revealed. That essence expresses the wholeness and personality of the teacher but also transcends the teacher. That feeling of "something more" – the emergence of newness from the interaction of teacher, student, and the words on the page – Rabbi Shapiro calls prophecy, not in the sense of a miraculous telling of the future, but in the sense that the student becomes "enthused" (filled with God) and they themselves become a creative wellspring of Torah.

These qualities that Rabbi Shapiro describes are also what make the natural world alive and vibrant. The sense of spirituality that we get in nature emerges from the wholeness of its myriad patterns. Emergence—the new, unpredictable properties of the whole that go beyond the properties of the parts—is found in every ecosystem, small farm or community. And when we see the fences and fields, orchards and gardens of a small farm creating a balance and a dynamic pattern, we sense something mysteriously added that we label the unique beauty or natural charm of a place.

What we see in Rabbi Shapiro's description of learning Torah is one example of the world emerging, a world in which the Divine can break through into our consciousness, be it through learning a Torah text or composting in our backyard. Suddenly, without warning, we can be struck with an intimation of the wholeness that is both beyond and within this place, time or person. Call it spirituality, prophecy, inspiration or simply God – emergence feels holy.

Emergence as Craft
and Way of Life

The patterns of a healthy farm as Berry describes it are both efficient and beautiful. There is something intrinsically attractive about patterns, and that is a secret of emergence. In our world, beauty has often been trivialized. Philosopher and environmentalist Sandra Lubarsky has pointed out how beauty was once a central concern of philosophy, but has since been demoted. The arts are the first thing to be dropped when the budget gets tight. After all, it is "only aesthetics" and that is not the essential, brass tacks of getting the job done. On the scale of human history this attitude is fairly new. We can trace it back to the Scientific Revolution of the 17th century and the writings of people like Descartes who extolled objectivity and mastery over nature. Anything "subjective," such as beauty, became suspect and not as valuable as the objective measurements that science can give us.[33]

Yet, aesthetic judgments, the second-to-second feedback loops that run between the farmer and their field, cattle, soil and community, are the basis of a well-run, healthy farm. This is true in practically any field you can name. For example, Orit Kent, an Jewish educational theorist who has written on the traditional study method of *havruta* (study in partners), has developed a methodology of looking at teachers' "moves" within a classroom situation. These "moves" are the crucial moment-to-moment decisions such as when to jump in with a comment, what tone of voice to use when making a comment, how to affirm a student's

---

33 See Sandra Lubarsky, "Toward a Theology of Beauty: *Hiddur Mitzvah* as an Eco-theological Imperative," in *The Mountains Shall Drip Wine: Jews and the Environment* (Omaha: Creighton UP, 2007) ed. by Leonard Greenspoon, 73-86.

idea or make a correction. She says that a classroom is like a dance.[34] This aesthetic, moment-to-moment relationship with one's material, be it soil, canvas, economy or classroom, is the basis not only of decorative craft items, but carries the secret of a healthy and efficient society.

Challah and Halakhah

I'm a pretty good challah maker. Actually, I'm being too modest. I make some of the best whole wheat challah most people have ever tasted. I think part of the reason it comes out so well is that at crucial points in the process, I don't rely on measurements but on my senses to see what is happening with the dough. I add flour to the liquid little by little, so that I get the exact right consistency for the dough. I can feel the weight, the bounciness, the shine of the dough. I know in my hands when it is ready. No measurement in a recipe could get that kind of accuracy. If you simply follow measurements in a recipe, you don't know all sorts of things such as the exact coarseness of the flour, the temperature that day, the hardness of the water, etc. When you knead and feel the dough, you get exactly the consistency you want.

So, I was happy to see that one of my favorite writers on food, Michael Pollan, is on the same page. In his book *Cooked: A Natural History of Transformation,* Pollan at first goes into the world of bread baking with the normal expectations that this is a world for the exacting, perfectionist, carpenter or computer coder types. Not the artist, poet type with which he identifies. But that turns out not to be the case. Yes, he needed to buy a scale that measured in grams, but when he sat down to read the recipe for one of the most amazing breads he had ever tasted, the Tartine bread made

---

34 See Orit Kent and Eli Holzer, *A Philosophy of Havruta: Understanding and Teaching the Art of Text Study in Pairs.* (Brookline, MA, Academic Studies Press) *2014.* Also personal communication.

by Chad Robertson, he found that being exacting was not the goal. Pollan writes:

> Robertson encouraged bakers to be observant, flexible, and intuitive. … He talked about dough as if it was a living thing, local and particular and subject to so many contingencies that to generalize or make hard-and-fast rules for its management was impossible. Robertson seemed to be suggesting that success as a baker demanded a certain amount of negative capability – a willingness to exist amid uncertainty. His was a world of craft rather than engineering, one where "digital" referred exclusively to the fingers.[35]

It turns out that this intuitive approach is also the secret to Jewish civilization. For more than 1,500 years Jews have followed Jewish law, *halakhah* in Hebrew, a word that comes from the word "to walk." It is a path, an active moving, one step at a time. It is not some theory proposed by a philosopher in an ivory tower. It is not a recipe. It is, at least when it is working well, a constantly evolving, constantly fine-tuned relationship between lawyers/scholars/teachers (usually rabbis) and their community: everyone who is living a Jewish life, walking on the path.

In *halakhah*, there is no one set of principles or abstract deductions on what is Right or Truth. Rather, there are stories, examples and arguments all woven together. Any rabbi or knowledgeable individual who would decide Jewish law needs to have a feel for the law and a feel for the individual people involved. Then a decision can come out exactly right for that moment.

Of course, it didn't always, and doesn't always, work that way. But apparently it has worked that way often enough that the Jewish people survived for 2,000 years in difficult exile and are miraculously still here today. The organic structure of learning,

---

35 Michael Pollan, *Cooked: A Natural History of Transformation*, (New York, Penguin) 2013, 215.

action and law has been flexible and durable, like an ecosystem adjusting to new conditions and continuing to grow.

In the case of making challah, the baker has to develop a sensitivity to the dough. When I'm kneading the dough, I need to be in second-to-second contact, to feel in my hands the dough's weight, stickiness, and bounce. It's a feedback loop, as the dough responds to my hands and my hands respond to the dough. When the dough and my hands become interdependent—each intimately responding to the other—then something unique can emerge.

In the case of an organically alive Jewish community, there is a dynamic reciprocity between the tradition and the lives of the individuals who uphold it. It's not limited to rabbis and big legal decisions. Any person's study of Talmud or other books of Jewish learning—if successful—is a back-and-forth interaction between the text and our values and insights. Actually, it is often still more complex because the texts are traditionally studied with a *hevruta* (a study partner). So, the conversation is between me, my study partner and the text. And the Talmudic text is itself a record of centuries of such conversations. The wisdom or truth that one might look for in these texts is therefore not to be found preserved "in the book." It is a dynamic, emergent truth that evolves as the conversations evolve. I'm reminded of education theorist Parker Palmer's definition of truth as an "eternal conversation about things that matter, conducted with passion and discipline."[36]

The beauty of baking bread, or of any craft, is that while you are in intimate interaction with the materials, it is never possible to completely control them. You work hard to perfect your art; you have discipline and consistent practice to build your skills in order to create something beautiful. But so much of the beauty actually comes from the surprising things that happen in the interactions,and which can't be planned. The accomplished artist or craftsperson knows how to improvise and play off the

---

36 Parker Palmer, *The Courage to Teach: Exploring the Inner Landscape of a Teacher's Life* (San Francisco, John Wiley and Sons), 1998, 106.

aliveness of the materials, their fellow creators and their emerging creations themselves.

I think about this when I walk on the historic breakwater in Rockland, Maine. I always marvel at the way the huge slabs of granite were placed there, 700,000 tons of them lined up almost perfectly in a line on one side and more randomly on the other side. You see the record of a process. Someone back in the 1890s was making aesthetic decisions, planning where to place each granite slab, but also improvising and reacting as each stone was put into place. There is a kind of vitality in this record. It's the beauty of all craft; a feedback loop of action and reaction, a living system. It's a record of the interface between our striving for order and control and the dynamic randomness of the living world.

Whether it's me baking challah or a 19th-century engineer building a rock wall or an observant Jew following *halakha* and trying to discern God's will for them in this moment, the key is always to stay awake and alive to the ever-evolving present: using our skills to shape and form, but always open to feedback, looking, listening and feeling. In that sacred moment of aliveness we're always on the path. That sense of aliveness comes from what modern theorists have called Emergence and what I've nicknamed "minyan"—the sacred community that emerges when ten Jews gather. But how, exactly, does a living community emerge?

The Invisible Forces
that Make Us

In the world of Jewish life, texts, people and actions are woven together. The Torah and the Talmud become sources for human action, which changes people and society and influences more texts in a dynamic feedback loop. In Jewish life it is sometimes hard to know where the text ends and life begins. The Baal Shem Tov had a saying that every Jew represents a letter in the Torah. We are the Torah and the Torah is us, in many complex ways.

Rabbi Jonathan Sacks has a wonderful essay on the Baal Shem Tov's saying. He contrasts the ways we think of a fulfilling life in the modern secular world to this idea of every Jew being a letter in the Torah:

> We can see life as a succession of moments spent, like coins, in return for pleasure of various kinds. Or we can see our life as though it were a letter of the alphabet. A letter on its own has no meaning, yet when letters are joined to others they make words, words combine with others to make a sentence, sentences connect to make a paragraph, and paragraphs join to make a story. That is how the Baal Shem Tov understood life. Every Jew is a letter. Each Jewish family is a word, every community a sentence, and the Jewish people at any one time are a paragraph. The Jewish people though time constitute a story. ...[37]

---

37 Jonathan Sacks, *A Letter in the Scroll: Understanding our Jewish Identity and Exploring the Legacy of the World's Oldest Religion,* (New York, The Free Press), 2004.

Rabbi Sacks' poetic description captures the emergent quality of what we call covenant: a rather important word in Jewish life! Covenant stands as a centerpiece of Jewish theology. God entered into a covenant, eternally binding God with the Jewish people. Actually, it happened several times: first with Abraham, then with the entire Jewish people at Mount Sinai, then reaffirmed again a few more times for good measure. But from the perspective of emergence, covenant stands for relationship. Mutual obligations imply listening and responding, feeling a part of something larger than one's self, being a part of a whole. The experience of being part of a minyan opens the door to feeling a part of the larger entity, the Jewish people. It is with that larger entity, spanning time and space, that God makes a covenant.

Living a life as part of a covenant can be a completely different experience than the commodity analogy that Sacks indicates as our dominant contemporary metaphor. Our lives gain meaning in the larger context of the covenant. Meaning emerges from being a part of a larger whole.

Twentieth-century Western thought has come up with some useful concepts in this regard. The anthropological concepts of culture and society describe a class of phenomena that emerge when people come together. The social-scientific definition of culture is something like the beliefs, behaviors, objects, and other characteristics common to members of a particular group or society. Culture includes language, customs, values, norms, mores, rules, tools, technologies, products, organizations, and institutions. Closely related to the concept of culture is society. Basically, society is defined as a group of people who are living together in some organized way.

Society is still a contested term in some circles. People who want to argue for more self-reliance sometimes downplay the concept of society, saying in essence, "I see people, I don't see society." Former British Prime Minister Margaret Thatcher famously claimed, "There is no such thing as society." It's even more common to hear that kind of thing on this side of the

Atlantic; culture and society are not easy concepts for individualistic "make my own way" Americans to grasp. Yet, for anthropologists, sociologists and others who make use of the concepts of culture and society, the claim that anyone is the sole author of their life is nonsensical. We are social animals so thoroughly that our very emotions, our speech, even our bodies are shaped by the invisible, emergent, forces of culture and society.

I'm reminded of this when I go to my in-laws' house in Lenox, Massachusetts. At the top of the stairwell there is a wonderful photo of an ultra-Orthodox teenage boy and his mother waiting at a bus stop.[38] The son is not only very thin, but he is slouching, leaning against the wall in a pose that makes it look as if he has no muscles at all. He looks a bit bored, but not unhappy. His mother stands erect, her arms folded in front of her. She looks stern, alert, no nonsense. For anyone who has experience with the Haredi (ultra-Orthodox) world, these poses are no surprise: They fit the gender and societal roles as defined and reinforced in that society. Young men are encouraged to study. Athleticism is not only not valued, it is discouraged.

In our Western culture, this young man could have been the model for the old cartoons of the skinny kid who gets sand kicked in his face while the bully gets the beautiful girl. Yet, in his Ulta-Orthodox world, this young man conforms well to gender expectations. He won't have trouble finding a good *shiddukh* (match). Married women, on the other hand, are often the managers of large families. Though modest in dress, they are not wilting flowers in terms of getting things done and taking charge of the daily material challenges of life on their husband's sometimes meager stipend from the yeshivah.

I may be reading a lot into this photo. But that is what a good photographer wants, to evoke a thousand words of interpretation and imagination. And I don't think I'm far off. Our bodies, how we hold them, their very shape and size are just as much products

---

38 This photo is by Hank Paper. See www.hankpaper.com

of our societies as they are of our genetics. Culture is in us and all around us. We are formed by it and we form it. Culture is a classic example of an emergent phenomenon in that it is a property of the group. It manifests even in the bodies of individuals, but it can't be found under any microscope. Imagine if that young man got fed up with his ultra-Orthodox life and decided to drop it all, move to California and take up surfing. In a relatively short amount of time we could expect his body language to change dramatically. Culture forms us and we form culture in a constant back-and-forth feedback loop.

A Genuine Culture?

As I always do on my visits back to Honolulu to see my mother and other family (who still live at or near the old homestead), I recently attended Shabbat morning services at my mother's congregation, Sof Ma'arav. It's a wonderful, eclectic mix of people. Among them is an interesting, pleasant fellow, Alex Golub, an anthropologist who turns out to be a graduate of the same anthropology department and a student of the same brilliant, fear-inspiring advisor I worked under as an undergraduate at Reed College.

Sof Ma'arav is a havurah, which means it is simply a group of (mostly) Jews who get together to pray, study, eat and socialize. This group started in the 1970s, when the "Havurah movement" was popular. The term that is popular for almost the same thing today, that is, independent Jewish prayer communities that are lay-led and organized without a rabbi, is ... guess what: minyan.[39]

This Hawaiian havurah or minyan has no rabbi. They rent space in the Unitarian Universalist Church (which, by the way, the

---

39 The "Independent Minyan" movement has many parallels and also differences to the "Havurah movement" of the 1970s. For a comparison between the two see: http://www.myjewishlearning.com/practices/Ritual/Prayer/Synagogue_and_Religious_Leaders/independent-minyan.shtml

young Barack Obama attended), and they take turns "giving the drash"—a short interpretation of the Torah reading given after the Torah portion has been read from the scroll. On my visit Alex gave an inspiring drash. I especially appreciated it because he quoted from one of the greats of anthropology, Edward Sapir, in a 1924 paper that I had never heard of before, "Culture: Genuine and Spurious." In this remarkable paper Sapir argues that there are such things as "spurious" cultures: fragmented, shallow cultures in which the individual doesn't feel a personal stake in the whole, and where people pay lip service to their ideals and beliefs, but don't really believe the things they claim to believe. He was basically talking about the American culture that he was a part of.

Then there are "genuine" cultures, in which individuals feel they are not like cogs in a machine, but have a fulfilling, meaningful part to play. The elements of the culture fit together such that a coherent and satisfying whole emerges, and, whatever the core beliefs may be, people actually believe them. Of the things Sapir wrote about a genuine culture, Alex quoted this line:

> ... it reaches its greatest heights in comparatively small, autonomous groups. In fact, it is doubtful if a genuine culture ever properly belongs to more than such a restricted group, a group between the members of which there can be said to be something like direct, intensive spiritual contact. This direct contact is enriched by the common cultural heritage on which the minds of all are fed...[40]

As he said this, I realized that he was describing something of a "holy grail" (to mix in some mythology from Christianity) of what the Jewish world and many other groups have been looking for— a key to a self-sustaining, vibrant community.

---

[40] Edward Sapir, "Culture: Genuine and Spurious," *The American Journal of Sociology Vol. 19:4, January 1924, 401-29.*

The *Pearl* and the *Flame*

"Direct, intensive spiritual contact"—remember, this is 1924 and Sapir isn't using the word "spiritual" in the same way we use it today. He's not talking about closing eyes and meditating together. In his lexicon, "direct spiritual contact" means something like real, meaningful contact between people in ways that touch their true values and deepest sense of themselves. It is contact between people in a community that facilitates people expressing their values and their individual talents in meaningful communal action. They are valued contributors and feel themselves in alignment with their actions.

To explain "enriched by the common cultural heritage on which the minds are fed," Sapir uses the metaphor of a tree planted in good, rich soil, as opposed to thin, sandy soil. A strong, healthy culture is going to draw from deep, rich resources—but as important, it will rework and reimagine those deep resources so that they become its own. A living culture emerges when each person is a builder and not simply a consumer. That means people jumping in as participants. It means some kind of do-it-yourself Judaism, whatever that may mean for each person and community. It doesn't mean that learned leaders aren't needed, nor that everyone is a rabbi. It does mean that people should strive to join, to take some active role, in the centuries-long conversation which is Torah and the drama of Jewish life.

I sometimes hear people ask, a bit defensively, "Am I not a good Jew?" They are proud to be Jewish and are living good, ethical lives. They are even contributing in many ways to the Jewish community: going to services occasionally, volunteering, giving tzedakah, sending their kids to Hebrew school. But the "Jewish culture" that they are attached to is no longer genuine in the way Sapir was talking about. It is no longer alive. It holds the power of tradition, nostalgia, and a good dose of guilt, but it is not alive and growing. And in that context, asking, "Am I a good Jew?" is really the wrong question. What we really need to ask is: "Are we creating a living, breathing, Jewish culture in our community?"

That question gets us thinking about the things that Edward Sapir was thinking about: Is there intensive, direct contact within the group? Are we creating contexts where people can express their real selves, where they can express their talents and interests? Are we creating contexts where we create real bonds within the community? Do people really believe what we are saying in synagogue, or is it lip service? Are people acquiring the tools to be able to draw from the rich wellsprings of the tradition and make it their own?

These are all questions about the conditions for emergence. Like the first stirrings of life in the primordial soup, there needs to be the right size membrane, boundary or context, so that there are "chemical" reactions in the community: energy and feedback that keep the metabolism of the group healthy. That might mean not too big or too small. The Kavanna Jewish Coop in Seattle has consciously adopted what Malcolm Gladwell called the "Rule of 150," which said that a tribe is about 150 people. That is the number of people who can get to know one another by name and interact effectively. Gladwell offers various examples of corporations that use that rule and split their campuses to comply with it to great results. The Kavanna community decided that it would forgo the "normal" trajectory of a successful new synagogue (they don't call themselves a synagogue, but a co-op and are organized as a co-op), which is to grow and grow until you get to 500, 700, 1000 members. They said, no, when we get to 150 we're splitting.

There also needs to be a dynamic balance in the relationship between individuals and the textual tradition: not giving the tradition so much weight that it overburdens the individuals, burying their sense of authenticity and autonomy; but, also not giving our feelings and biases so much weight that we are closed to learning new insights from the tradition.

That dynamic balance relies on the art of listening to feedback, making constant small adjustments, being awake to short-term goals and the big picture, and making intuitive decisions within that context. It relies on knowing that we are interdependent with those around us. Like any living thing, dynamic balance emerges in a self-organizing, almost mysterious process. Like any play or novel or painting or stone wall, its creation will involve a lot of sweat and frustration and long hours of slogging through, as well as moments of insight, delight and revelation. A healthy minyan is more than simply ten Jews together. It is an emerging community, a manifestation of something that is beautiful and awe-inspiring. And alive.

Earth, Climate Change and the Emergence of God

As I write this, it's February and about 10 degrees Fahrenheit here in Boston. It's colder than usual for Boston, but in the upper Midwest it's 15 below with wind-chills up to 65 degrees below zero. The know-nothings who prefer rhetoric to understanding will say, "Ha! Where's your 'global warming'? Look how cold it is!" That has about as much validity as someone who sees the distended stomachs of starving children in Yemen and says, "What starvation? Their stomachs look full to me!"

That's an obscene extreme of misunderstanding, but the fact is that climate change has been difficult to understand for many people because it's more complex than, say, acid rain, or rivers catching fire, or eagles eating DDT and dying. Climate change is by its very nature a crisis of the system, not of individual species, resources or regions. Even when a huge hurricane hits, for a long time the media would be very reticent about connecting it to climate change. The experts will always come in with qualified statements like "one can't directly connect any one weather event to climate change, but...."

It's that "but..." that we need to start to understand. Presently, understanding complex phenomena is generally seen as the specialty of scientists, engineers and theoreticians. But it doesn't need to be that way. Traditional people often have a more intuitive grasp of organically connected, complex systems even if they don't talk about them in technical, scientific language. Whether we start from modern, scientific language or traditional organic thinking, we need to come around to this organic way of thinking. Our planet depends on it.

The *Pearl* and the *Flame*

The Five Waves

Rupert Ross tells a story that illustrates this organic way of thinking. He's writing about how he learned to better understand Indigenous systems of justice:

> I remember...speaking with an Inuit woman in Yellowknife. I had been talking about the insistence of Aboriginal people that the justice system look beyond the particular crime and try to examine all the events and forces that lead up to it. She immediately told me about something that her grandfather told her when he took her down to the shores of the Hudson Bay as a youngster. It took me a while to understand that she was talking about the same thing I was!
>
> Her grandfather told her that before she ventured out she had to learn how to look for and understand how the "five waves" were coming together on any particular day. . . . The first waves were those of the winds that were building up but not yet fully arrived, the waves that would grow strong as a new weather system came in. The second waves were the ones left over from the weather system that was now fading, for they would still continue to affect the water even after the winds had gone. The third were the waves caused by all the ocean currents that came winding around the points and over the shoals, for they would present their own forces against the waves from the winds. Fourth were the waves caused by what Westerners call the Gulf Stream, and fifth were the waves caused by the rotation of the earth. Until you looked out and saw how all those forces were coming together, then developed some idea of how they would interact as the day progressed, it was not safe to go out and mingle with them.

As I have slowly come to appreciate, the teaching of the five waves has direct application to the ways in which a justice system ought to approach offenders. It suggests, for instance, that we cannot come to understand their behavior until we gain some understanding of all the waves, old and new, that have converged on them during their lives. It suggests that they will continue to face the same waves tomorrow and the day after that, and that many of them cannot be changed. It suggests that what is necessary in the face of that reality is a process that helps them develop the skills they will need to ride all those waves more successfully in the future. Further, it suggests that the very *last* thing any justice process should do is cause a reduction in whatever riding skills offenders already possess.[41]

Ross is saying that he learned to appreciate an entirely different way of thinking about problems of justice and their solutions from the Inuit woman and from the Aboriginal justice systems. Yet, two significant things jump out at me: first, it is different, but not so different that he couldn't grasp it and come to embrace it. The mindset of seeing systems is latent in our minds and in our cultural backgrounds. We know that we are part of something larger, and we resonate and even yearn for a return to that way of thinking, even as another part of our mind resists it.

Second: this way of thinking applies across the board. Whether we're talking about a justice system, venturing out on the Hudson Bay, an activist organization or our planetary climate, we can benefit from shifting our perspective to see how the parts interact to create a whole system. If using the analogy of ocean currents doesn't work for you, we have modern, scientific ways of

41Rupert Ross, *Returning to the Teachings: Exploring Aboriginal Justice* (Toronto, Penguin Canada), 2006, 75-76.

talking about this way of thinking. It doesn't really matter where you start, but making the shift is essential.

A Fancy, Scientific Word: Stochastic

The term "stochastic" refers to a process where the individual events can't be precisely predicted, but the system is following a predictable pattern. The fact is that the hurricanes, droughts and floods and fires are entirely predictable in terms of the complex feedback loops that make up the global climate systems. When the oceans are warmer, more moisture evaporates into the air, which causes more severe storms. The warm air also dries out the trees inland, making more fires predictable. But you can't tell exactly where and when these individual events are going to hit. Any complex system works this way. When you throw a rock at a glass window at enough speed, the glass will break. But where exactly will the lines of fracture go? Probably impossible to know. When you heat up water until it gets to 212 degrees Fahrenheit it will boil, you just don't know which molecule will start the chain reaction. When you heat up the planet enough, its systems fall apart. That's a fact. Exactly where and when each storm, each drought or fire will hit isn't precisely predictable.

And, the nature of feedback loops is that they can sometimes become what is called "positive feedback" which often isn't "positive" in the sense of "beneficial" but rather, it is a runaway cycle, a "snowball effect." We've seen a lot of that in the world of climate change. Even scientists using the best computer models seem to be continually surprised at how fast the change is happening. That's because of underestimating those positive feedback loops. For example, the methane gas that is trapped under the arctic tundra is a potent greenhouse gas. When the tundra heats up, it releases the methane, which creates more heat, which feeds back and releases even more methane, and so on.

It's not only in climate change that we need to start understanding these stochastic processes. My friend Jed

Shugerman, a law professor and political blogger, has been using the term "stochastic terrorism" regarding the rise in racist, white supremacist violence in this country. While it is highly unlikely that any one act of terror can be exactly causally traced to any one speech or tweet by a powerful political figure, the trend is clear. The rhetoric of political leaders does create the system-wide conditions for more domestic terrorism.

It is the same dynamic with gun violence. If we're looking at the individual incident, a shooting of innocent people, the first instinct is to say, "If there was someone with a gun there, they could have stopped it more quickly." But, looking at the larger picture, we know that the more guns we have out in people's hands, the more violence, the more deaths. The countries that have lessened their guns have lessened their gun violence. Even the U.S. states that have enacted gun laws have lessened their gun violence. But that doesn't convince people. Our modern, Western minds are trained to look at the individual incident, and to miss the larger, system wide dynamics.

So, when we approach climate change, we are in a different ball game than we were, say, back in the 1970s when we were fighting water pollution, air pollution, and saving the whales. In those "good old days" of environmental action, we could at least picture the cause and effect; we could identify a clear and often photogenic victim. But with climate change, there are often system-wide effects, the nexus of causation is complicated and indirect. And because it is system-wide the crisis is on another scale of disastrous. Our whole planetary system is fracturing.

We need to start connecting the dots and getting in the habits of seeing the whole, interacting system, or system of systems. It's another mind set. Another way of being in the world.

The Spirit of a Place

Rebbe Nachman of Breslav [1772 – 1810], the Hasidic mystic and great-grandson of the Baal Shem Tov, had finally made the epic journey from his home in Eastern Europe to the Land of Israel. When he arrived there he found, as all people who travel there do, that the Holy Land of Israel is made up of rocks, dirt, trees and plants—just like other places.

> But in the imagination of all those generations of exiled Jews and based on all the glowing descriptions in the holy books, the Land of Israel is supposed to exist on an entirely different plane of spiritual existence. The Zohar says, for example, that the Land of Israel has its own sky, separate from the skies of all other countries. Rebbe Nachman was not fooled by the normal looking land. He experienced, even with its normal rocks and trees and plants, the special holiness to the Land of Israel. Walking even a few steps there was a high point of his life. (adapted from Likutey Moharan, Tanina, #116.)

Rebbe Nachman's experience of the holiness of the Land of Israel is only one example of a widespread phenomenon: there is a mysterious way that one can sense a difference in a place not in any one detail, but something that arises out of the whole. There is an essence of "The Land of Israel" that is not seen in any stone or tree, just as the Piaseczner Rebbe taught in our example above, that a teacher or a book can transmit something in their wholeness that you can't point out in any one detail. Rebbe Nachman's experience reflected his mystical, kabbalistic background, but there is a simple statement from the earlier sages of the Talmud that, in its own way, makes a similar point: "The air of the Land of Israel makes one wise." [Babylonian Talmud, Bava Batra 158b].

Air is a perfect image to convey the sense of something invisible that floats above, surrounding and enveloping a place. I think that the Talmudic rabbis were saying essentially the same thing as Rebbe Nachman: there is something that hovers around a place, invisible yet very real, that changes everything. Perhaps Israel does have its own "sky," or perhaps every place has its own sky: a hovering sense of the uniqueness of each place when you grasp it as a whole. Biblical scholar Ellen F. Davis quotes Tim Gorringe:

> There are spiritualities which give rise to Cotswold villages, Italian hill towns, compact Indian cities, to the Parthenon, the Dome of the Rock, to Chartres, and the Meenakshi temple, and there are spiritualities which give rise to the unsustainable cities of Assyria and Babylon, to slums, the enclosing of rivers, Disneyworld, and the carving up of cities by grotesquely misnamed "freeways."[42]

One can say that a spark of God emerges in the "spiritualities" that Gorringe writes about. One level of the emergence of God happens through particular places, through ecosystems that are alive, that have what Bateson called mind. On the larger scale, Gaia, the planetary life of earth, emerges in the world through the combination of these "spiritualities" and when we destroy the patterns and complex feedback loops that make the world healthy/whole/holy we damage God's presence and manifestation in the world.

Yet, the deeply woven spiritualities of the earth are resilient. There is hope in the power and depth of these ancient, living systems. One of the most moving things that I have read on this subject is an interview with the late marine biologist, Eva Saulitis. She worked with the orcas in the Prince Williams Sound and came to know and love the populations of orcas that swam in these waters, even as they were becoming extinct. Her life

---

42 Ellen F. Davis, *Scripture, Culture and Agriculture: An Agrarian Reading of the Bible* (Cambridge, Cambridge University Press), 2009, p. 107, quoting T. J. Gorringe, *A Theology of the Built Environment: Justice, Empowerment, Redemption* (Cambridge, Cambridge University Press), 2002, 243-44.

tragically paralleled the orcas, with her cancer diagnosis and finally, her death. Yet, she maintained a strong sense of hope. It was a hope that was at once daring and idealistic and also clear eyed and realistic. She writes:

> There's stubbornness in that prayer that you read from the end of the book [*Into Great Silence*, Eva Saulitas, Beacon Press, 2013], "that what's broken can be mended. That what's shattered can be made whole. That what's damaged can be repaired." The Chugach transients (a population of orcas in the Prince Williams Sound) will be lost to this earth, and yet the Sound is a place where healing continually occurs. Yes, it's a threatened place. Climate change and ocean acidification and loss of salmon are all grave threats. And yet at the center of that place is a palpable force of healing that is unstoppable. It just is.
>
> The Eyak are a people of Prince William Sound whose language is now extinct in native speakers. Some of the last speakers of Eyak insisted that their language was embedded in the place, and as long as the place remained, their language would come back. [Voice breaks.] These people, whose language is gone, whose culture is gone, who have lived in that place for ten thousand years or more—they know what they are talking about. I believe there's a truth in what they're saying, something we don't understand about how the earth heals, what losses mean to a place, how it's all entwined. The Eyak fiercely believe that there will be people in the future who will come to know the land in such a way that they will re-create the Eyak language. It will arise again. Maybe it won't be exactly

the same. Maybe it will be in another form. But the land, the earth is the center.[43]

I find no better example of the spirit of a place than the fact that a language can grow out of that place and its unique pattern of geography, plants, animals, water, sun and air. Eva Saulitis's hope and trust in that spirit is deeply religious in a non-sectarian register. I'll be coming back to this question of hope and faith in the section on Mitzvah/Tipping Points, but for now, it is a testament to the emergence of spirit; the depth and strength of that emergence which can withstand the destructiveness of our contemporary greed, shortsightedness and arrogance.

The Breathing Earth and the Name of God

Rabbi Arthur Waskow expresses a no less hopeful, but more activist, passion about the state of the planet. He has written and spoken about Climate Change as a crisis in the Name of God. Since the holy, unpronounceable Name of God in Hebrew is made up entirely of vowels Waskow explains, it is essentially breath. In Hebrew the letters, Yud, Heh, Vav, Heh, which spell the name of God are a grammatically impossible mixture of the tenses of the verb "To Be." God is Being in the past, present and future all at once. They are all vowels and their pronunciation is all breath.

If we don't understand God as somehow standing outside the world, but instead infusing and animating the world, then the earth, with its balance of breath from plants and trees breathing in CO_2 and breathing out oxygen, while the animals (including us) breathe in oxygen and breathe out CO_2, is a manifestation of the name of God. And the Name of God is in trouble. The balance has been disturbed and the flow of blessing and flourishing that this planet could provide is turning into the curses of storms, fires,

---

43 From an interview with Eva Saulitis by Christine Byl in *The Sun Magazine*, January, 2017.

flood and drought. The wholeness/holiness of this earth has been profaned, and we are seeing the result, not in "supernatural" punishments from a Sunday school imagination, but with real, natural and yet equally divine consequences of our greed and short-sightedness.[44]

Connectedness versus Addiction

That aliveness that comes in our flow of interactions plays a bigger part in our health and well-being than we're usually aware of. Despite the mythology of our culture that raises the individual above these connections and makes it seem that we exist on our own much more than we actually do, we humans are built, physically, emotionally and spiritually, to be a part of something.

The crisis that we are currently experiencing in the United States and other parts of the world with addiction has many complex roots. But in the public discourse I hear a lot of what feels to me groping around in the trees and bushes rather than seeing the forest. We've made a lot of progress—It's much less common today to hear about addiction as a moral failing than it used to be. Although there's still a long way to go, there is a strong and growing movement away from criminalizing drug addiction. Yet, the alternative is most often to say that addiction is a disease. That is also true, but it still focuses on the individual. It doesn't answer the obvious question: why, all of a sudden, did this disease become an epidemic in developed countries? Why wasn't it a common disease in other societies?

There has been good research that argues that we need to look to the roots of the addiction crisis in our societies and the

---

44 See for the latest of his many published works, *Dancing in God's Earthquake: The Coming Transformation of Religion,* (Maryknoll, N.Y., Orbis Books), 2020. See also the excellent book by David Seidenberg, *Kabbalah and Ecology: God's Image in the More-Than-Human World,* (New York, Cambridge University Press), 2015, for many sources from kabbalah and midrash which place the manifestations of God, the Divine Image, in the wholeness of the natural world.

fragmenting of coherence and connectedness that is the root of so many of our problems. Bruce Alexander, a researcher at Simon Fraser University in British Columbia, has been the leader in this approach to addiction. In the 1970s, he found he was dissatisfied with the research being done on rats in connection to opioid addiction. The rats would press levers to get cocaine and would quickly become so addicted that they would ignore pressing levers for food, preferring the drug.

But, Alexander reasoned, these rats are severely deprived of their natural living conditions. They are social animals. So, he created what his lab called the "Rat Park." It was filled with wheels and shavings and tunnels and plenty of other rats to play with. In the "Rat Park" the rats tried cocaine but they didn't get addicted to it.

Humans as well are much different when we are in a healthy, connected community. Being a part of something larger fulfills for us something essential. Like so many things, we can trace our present malaise to the disruption of our communities and our sense of being a part of something. Alexander himself puts the problems of addiction, spirituality and climate change together: the root problem is our losing that sense of being a part of something larger. Our addiction problems are not just with drugs, but with a consumer culture that uses too much energy, but doesn't really bring satisfaction. Here is an excerpt from an interview with him in *The Sun Magazine:*

> The social and political system past generations struggled to create has been twisted into a cruel and stupid imperial system dominated by multinational corporations. This is hard for people to admit. Who can face the fact that the consumer society we were raised to cherish is actually making us apathetic, crazy and vulnerable to addiction? The disconnected, fragmented nature of our culture causes addiction, which causes further fragmentation...

The *Pearl* and the *Flame*

Lack of belonging is a feeling of being alone and neglected. Lack of identity is the unease of not knowing who you are and experiencing wild swings from one self-concept to another. Lack of meaning is a sense that the world is random or ruled by evil forces. Lack of purpose is boredom and a feeling of uselessness, of not having any reason to get out of bed. When all four of these are unsatisfied, life is hell. . . The process of weaving together belonging, identity, meaning, and purpose is usually accomplished through a living culture, which we might say includes a mysterious spiritual component. We know that in cultures where everyone has a place and a purpose and a stable way of life, addiction is very rare. ...our ecological problems and our addiction problems are linked.[45]

Judaism has also connected addiction to this longing for our true selves, and ultimately, to God. Any addiction is a perversion of the true yearnings of our soul according to the Rabbi Kalonymous Kalman Shapiro, The Piaseczner Rebbe. He writes:

There are two kinds of work in the world: work of service and work of worry. And it all depends on what a person wants to get out of the world. One who wants to get worldliness and materiality...all their lives will be filled with anger and discontent. It's not even that this person necessarily has a lot of troubles, rather, more generally, that person's inner self isn't ever satisfied, and they don't get satisfaction from the world. So, they are constantly seeking to calm their suffering soul—these are the desires that the person feverishly seeks in order to satisfy the pain of their

45 "Filling the Void: Bruce Alexander on How our Culture is Making us Addicted" by Jari Chevalier, *The Sun Magazine*, March 2019.

soul. It can be compared to a person who smells a bad smell – they look for some spices or other good smelling things in order to lessen the suffering of their nostrils because of the bad smell. This is not the case for the person who doesn't smell anything bad, (that person) doesn't need to search for spices. But the God Wrestler,[46] who knows that it wasn't for the matters of materiality that they were sent into this world, and who constantly remembers that in the short time that they have in this world, they try not to lose their connection to eternity and to the upper worlds—and that is it rather to learn Torah and to serve that they came into this world. . . that person will be above the toil and exhaustion.[47]

If we find God in community, in the earth, and in our living soul, we feel the wholeness and joy of living in this world, even with its sorrows and pain. There is tragedy and love, life and loss, yet, when we are a part of a living community, when we feel immersed in a world that holds us and makes our lives meaningful, we can find strength and inner joy even in the hard times. God, in the highest heavens may be the ultimate Source of All, the Infinite Unknowable, yet, for us, in this world, God manifests in the life, strength and joy emerging from the wholeness of the patterns of life that we weave. When we lose touch with those living, God-filled patterns, we hurt ourselves, descending into addictions, and, in our many addictions, such as our addiction to energy guzzling consumer culture, we destroy the world as well.

---

46 "God Wrestler" is one way to translate the name Israel, which is the traditional name for Jews: the Children of Israel, or Israel for short. By translating Israel as God Wrestler we also leave open the option of a more universal understanding of the term. Someone who is wrestling with God, struggling to understand and come closer to God may also be a God Wrestler. This is not meant to erase the particularity of the Jewish people, but it opens up the option to say that others as well may be in the position of God Wrestler.
47 *Derekh HaMelekh*, Parshat Chayei Sara (1931).

PART THREE:

Mikdash

The Concentric Circles of Life

"There is one who sings the song of his soul, discovering in his soul everything—utter spiritual fulfillment. There is one who sings the song of his people. Emerging from the private circle of his soul—not expansive enough, not yet tranquil—he strives for fierce heights, clinging to the entire community of Israel in tender love. Together with her, he sings her song, feels her anguish, delights in her hopes. He conceives profound insights into her past and her future, deftly probing the inwardness of her spirit with the wisdom of love. There is one whose soul expands until it extends beyond the border of Israel, singing the song of humanity. In the glory of the entire human race, in the glory of the human form, his spirit spreads, aspiring to the goal of humankind, envisioning its consummation. From this spring of life, he draws all his deepest reflections, his searching, striving and vision. Then there is one who expands even further until he unites with all of existence, with all creatures, with all worlds, singing a song with them all. There is one who ascends with all these songs in unison—the song of the soul, the song of the nation, the song of humanity, the song of the cosmos—resounding together, blending in harmony, circulating the sap of life, the sound of holy joy."

— Rabbi Abraham Isaac Kook, *Orot Ha'Kodesh* 2:444-5 (translated in Daniel Matt, *The Essential Kabbalah*, 154).

"Complex systems also tend to be nested, one within another, and are separated by fuzzy boundaries. *These are boundaries that allow for the flow of energy, materials, and information between larger—and smaller-scale systems, but maintain each system's integrity. Biological and political systems demonstrate these attributes well."*

— Tom Wessels, *The Myth of Progress*, 12.

Sanctuaries, Coverings
and Boundaries

Ever since modernity we've had this crazy dilemma: we want our freedom, but we also want our old feelings of tribe, family, belonging. Let's face it: we've been trying for over one hundred years to solve this: socialism, capitalism, psychotherapy, communes, mass movements, mass media, fundamentalist religion, Twitter, opiates—spinning, pivoting back and forth between looking for belonging and reaching out for more freedom, more convenience, more individuality.

It feels like we're getting to a breaking point—political institutions crumbling before our eyes, neighbors living in alternative universes depending on which news channel they listen to. How do we find a way that helps us integrate both our modern values of freedom, individuality, universal rights and also belonging, family, tribe?

In my journey into Judaism and then, into ecology, I found a key, a different way of thinking about our place in the world that doesn't get hung up on the horns of that dilemma of universalism versus tribalism. I've come to call it *Mikdash:* Sanctuary, because a sanctuary, such as the Holy of Holies in the ancient Temple, or the sanctuary of the human heart of someone saying the Sh'ma, by its nature is nested, fractal, emanating through many levels. That Holy of Holies stood at the center, but it radiated holiness out into the courts, past the walls, out to Jerusalem and beyond. The heart of the God Wrestler saying Sh'ma points to the hands, the eyes, the doorposts, and the gates along the way out into the world. It's the same way that a cell is nested in an organ, which is nested in a body, which is nested in a family, and all the way on up and down the scale.

The *Pearl* and the *Flame*

I didn't discover this all at once, but back in my twenties, I made the discoveries that put me on the path of understanding that sometimes, dividing, covering, holding and creating boundaries can lead to more unity.

Doing Anthropology on Myself

That either/or dilemma that I experienced in my one day at the *"ba'al t'shuvah"* (aiming to bring back secular Jews to "Torah-True" Orthodox Judaism) Aish HaTorah yeshivah in Jerusalem when I was twenty made me believe I had to choose: between my universalist secular world, and the insular, spiritually vital world of yeshivah, but instead that false choice broke me apart. I made my escape from Aish HaTorah and headed back to Portland, Oregon, to attend Reed College. But even as I left the yeshivah, walking quickly down the alley, my insides started to shake, and I felt like I was caught in a vice, my breathing shallow as if I could barely connect with the surrounding air. My feet were leaving, but my mind and heart were still undecided.

How could I be returning to what seemed like the emptiness and moral neutrality of the wide secular world? I was returning to the same hyper-individualist and secular scene of Reed College that had pushed me toward exploring Israel in the first place—searching for the echoes of that warmth and feeling of community that I had tasted in Jewish summer camp in California. But, on the other hand, how could I fit myself into the small world of these black hatted, science denying, manipulative rabbis?

I made my way back to Reed and had an angst filled fall semester, spending a fair amount of time in the doctor's quiet room to calm my nerves, talking to (secular, Jewish) therapists who didn't have a clue as to what I was talking about, reading anything I could get my hands on about Judaism and talking to anyone who seemed to know something about Judaism.

The one person I spoke with who had a concrete idea for me was Rabbi Joshua Stampfer of the Conservative synagogue,

Neve Shalom, in Portland. He didn't try to solve my theological dilemmas, but suggested that I work that summer as a counselor at the small summer camp he helped to run up in Washington State. The summer camp was more traditional than I had grown up with, but not at all like the yeshivah in Jerusalem. I met people my age who were praying in the traditional Hebrew liturgy, but they also talked about sex and movies and other normal things. The guys wore the leather straps on their arms and head called tefillin during morning prayers, but they also wore jeans and tee shirts. They didn't cook food or turn on lights from Friday evening to Saturday at sundown, but they also worked for Soviet Jewry, went for long runs in the forest with their dogs, and threw the counselors into the lake for fun. I relaxed a little.

That summer experience gave me the courage to take what felt like a big step: leave Reed for the next fall semester and head back to Israel. It was the first time that I had stepped off the smooth, curated path that I had been on since I started kindergarten. I needed a psychic push to clear myself from the orbit of my friends and the expectations of graduating in four years. Little did I expect that one semester off would turn into two years and I would come back a different person. Or, maybe I did have an inkling, and that's why it felt so scary.

This time around I went to a secular kibbutz, Mishmar HaNegev, to learn Hebrew for a few months, saying to myself that I wanted to find my own path, get the tools to learn about Judaism, and not be taken in by these rabbis. I did begin to learn Hebrew and enjoyed some parts of kibbutz volunteer life: getting up early to work in the fields (although I was mostly in the chicken house which was less pleasant), learning how to eat the kibbutz breakfast cutting up my tomatoes, cucumbers, peppers and onions really small and mixing them up with rich, creamy white cheese. Toasting bread on the smelly kerosene heaters we had in our rooms. But after about three or four months, being essentially a spiritual seeker, I got restless with the routines of the kibbutz and started looking for something new.

The *Pearl* and the *Flame*

I had seen a brochure back at Reed College for another volunteer program at a place called Ramat HaNegev College that advertised a work/study program of ecology, art, and social science while volunteering in a "development town" called Yerucham. When I started getting bored with kibbutz, I looked them up and decided to switch. It wasn't far away as my kibbutz was already in the Negev, so it was a quick bus ride to Yerucham. When I followed the directions to the college, I literally walked right past it. For all the slick brochure advertising, Ramat HaNegev College was a little portable classroom plunked down on one of the dusty streets of the town. I later found out that even that was a mistake. The building was supposed to be for a special education classroom in one of the towns up in the north, but someone delivered it to the wrong town. Welcome to Israel, 1979.

But I enjoyed Yerucham. I extended my leave from school and I ended up spending another few months volunteering in this "development town"—so called because, in theory, they were meant to develop into thriving, or at least self-sustaining, towns. In reality, they were small, isolated and economically depressed cities stuck out in the middle of the Negev desert and other remote areas by the Ashkenazi (European Jewish) dominated government. It was a way to put the dark skinned North African Jewish immigrants somewhere other than the big Ashkenazi cities. The in-gathering of the exiles was a bit more complicated than we learned in Sunday school.

But I drew sketches and made paintings of the desert scenes, helped to build a model passive-solar adobe house, hung out in our apartment playing music with the other volunteers and the riffraff of the town who liked to hang out with the volunteers, coached a youth basketball team, learned Hebrew by using it in daily interactions and generally had a wonderful few months. I was also taken in under the wing of one of my neighbors in the apartment block, Ya'acov, a very intense, newly ultra-Orthodox, newly married guy from New Jersey, who invited me to join him, his wife and infant for Shabbat meals and talked with me about Torah.

As with the rabbis at Aish HaTorah, he said a lot of crazy sounding things I didn't believe, but there was power in his words because many of them have stayed with me to this day. He said, for example, that one shouldn't dig around in your pocket to see how much money you have there because you could be preventing a miracle from happening—someday you might put your hand into your pocket and find you have more than you thought—just enough for what you need. I didn't know it at the time, but he was referring to a line from the Talmud which I later learned: "Blessing isn't found except on that which is hidden from the eye." [*Babylonian Talmud*, Bava Metzia 42a]

I later came to understand the wisdom of this statement in my own way: When we obsess about controlling everything, "checking our pockets" all the time so that we know exactly where we stand, we tighten up and don't let ourselves enter the unpredictable give and take of relationship with the world around us. We cut ourselves off from miracles and blessings.

I think about this, for example, in the context of teaching: if I'm so afraid that I won't give enough material and structure to the class that I fill up every minute locked into my lesson plan, I end up choking off any new ideas and energy which might emerge when the students are given a chance to ponder, discuss and think about the material. I've learned that tolerating a few seconds of silence can be well worth it.

I think Ya'akov meant it literally—don't check your pockets because God might make a deposit—but somehow, magical statements like that, which seemed to come from another world, struck a chord in my soul and stayed with me until years later I sometimes came to see a more profound meaning there.

Ya'akov was pushing me to go back to one of the Ultra-Orthodox Yeshivot in Jerusalem, but I also knew some other folks in town who told me about a more modern yeshivah, one that integrated traditional text learning with a good dose of openness and modern liberal ideas. After some explorations and trying places on for size, I ended up enrolling in the Pardes Institute in Jerusalem and extending my leave of absence for a second year. It was, indeed, the beginning of my integration.

So, in my second year in Israel, I found myself sitting in a classroom on the second floor of a drafty stone building in the South East section of Jerusalem called Bak'a. As opposed to most Orthodox institutions, at Pardes men and women study together. The students didn't live in dorms like the other yeshivot, and we called the teachers by their first names. This felt like someplace where I could actually integrate my life, rather than swap it for a new one.

The teacher in this particular class, a brilliant young rabbi named Dov Berkovits, was talking about the Mishnah, an ancient Rabbinic text from the 3rd century C.E. And I was having a strange experience. The only way I've been able to describe it is that it felt to me like the rabbis who had composed this text were doing anthropology on themselves.

I was an anthropology major at Reed, and I was recognizing something familiar here in this Jerusalem classroom. On a surface level, we were supposedly studying the laws of reading "the Megillah," the scroll of Esther, on Purim. But these Mishnah texts were doing it in a very roundabout way, which I would come to learn is typical of these rabbinic writings. They were defining social groups such as towns, villages, walled cities, and their relationships with one another. Each social group read the scroll on different days, or sometimes the small villages combined with the larger towns on market days.

> The Megillah is read on the eleventh, the twelfth, the thirteenth, the fourteenth, and the fifteenth [of Adar], never earlier and never later. Cities which have been walled since the days of Joshua ben Nun read on the fifteenth; villages and large towns read on the

fourteenth; Except that villages move the reading up to the day of gathering (in the large cities for market day). (Mishnah Megillah 1:1)

Dov showed us how the patterns in the text suggested how sacred time was not some abstract idea, given on a silver platter from God, but was malleable, stretching and re-creating itself in relationship to our social world. The walls of an Israelite city, we learned, were deeply resonant symbols for the holiness of the people. The walls were a vessel, like a cell, that held and defined an area of holiness that seemed to emanate from the group itself.[48]

This was a modern Orthodox rabbi teaching a two-thousand-year-old sacred text, but I saw how this was the same kind of thinking that anthropologists like Emile Durkheim, one of the most influential figures in 20th century anthropology (and Jewish, by the way), used to understand society and religion. How we organize ourselves into social groups influences how we feel and think about everything: our bodies, our clothing, everything, including God.

As we studied this short chapter of the Mishnah, it became clear the rabbis had collected many items of biblical and rabbinic culture and arranged them in positions of juxtaposition and contrast: Festival Days or the Shabbat, vows or offerings, tefillin and mezuzot or books of the Torah, the cities of Jerusalem or Shiloh, small altars versus large public ones. Dov brilliantly showed us how they weren't put there just as mnemonic devices (as most of the academic scholars have posited), but they were telling a story. The story wasn't told in a narrative form, but was found in the juxtaposition and patterning of those items. The Priest and the leper (as we saw in Chapter Six) weren't randomly juxtaposed, but were, in their bodies and clothing, deeply

---

48 See the appendix for a still abbreviated, but more complete version of the analysis of this chapter of the Mishnah which I learned in that classroom in 1980, and later turned, with some changes and additions, into a chapter of my dissertation at U.C. Berkeley.

symbolic of patterns of control/spontaneity, life/death in ways that Durkheimian anthropologist Mary Douglas had written about. The rabbis of the Mishnah were following the same basic method as Ben Azzai in his stringing pearls: creating new juxtapositions and new meanings—new revelations.

I didn't know about Ben Azzai at the time, but all I could think was that these rabbis were somehow thinking about culture in the same way anthropologists do: by fitting bits and pieces together into patterns of inclusion, juxtaposition, boundaries and relationships. But they differed, of course, from anthropologists in that they weren't doing this as an academic, abstract search for knowledge of another culture. They were taking the fragments of their own past and putting them together to tell a new story, one that made sense of their new circumstance, but still connected them to that past. They were doing "anthropology" on their culture as a practical, survival practice: fitting together their broken world into something that made enough sense for them to continue moving forward.

I resonated with the way of thinking in the Mishnah, and it made me realize that my worlds didn't need to be compartmentalized. The either/or dilemma of big, universal, modern world, versus small, particular, Jewish world was melting into something much more subtle and dynamic. The smaller Jewish world was one part to be explored, but it connected to the larger world and could even play a role in helping that larger world. I saw how I could bring this Mishnah back to college and integrate it into my anthropology thesis. This was anthropology that I felt personally connected to. It was the integration that I've followed ever since. I no longer felt that I was floating, untethered and unrelated in a lonely, random expanse. I had a place to do my thing, an identity and community that grounded me, and I found I could connect to the rest of the world much more effectively when I knew who I was.

The next year, I returned to Reed College for my senior year. It was my best, most enjoyable year. I got together with a friend,

Miles, who had taken a similar journey to Israel and back, and we started one of the group-houses that Reedies often lived in. We found three other students, Susan, Deidre and Pam, who all had a relationship with Judaism of one kind or another, and we created a delightfully anomalous kosher, shabbat observant Reed house. I did a thesis in Anthropology of the Mishnah, but the anthropology was also on myself: I had begun piecing together the parts of my life into something that held together enough to make some sense. I was beginning to connect to my heart, to find a community, and with the strength that gave me, I began to find my way in the world.

Like the message of Mishnah Megillah, which Dov Berkovits had taught us in Jerusalem, I was finding that holiness isn't an abstract idea. I was finding it in community, in my household, in my college, in the Jewish people, and the larger world. Life was beginning to fit together. Looking back, I started to feel nested in the world. I was also beginning to grasp something that felt important, but I didn't yet have the words for it. This nestedness felt flowing and dynamic, alive. I hadn't quite put together that this fractal, nested way of patterning was one of the keys to all life—biological, social and ecological. And this living wisdom can be vital to dealing with the mess that we've made of our world.

The Complexity of Oneness

What is a sanctuary? We use it in speech all the time: "The Torah Service will be held in the main sanctuary at 11:00." Sanctuary (*mikdash*) is simply a container, a home for sanctity, holiness. How many books have been written about what sanctity, or holiness, is! For now, let's go back to that philological exploration that we saw from Wendell Berry: holiness is related to wholeness, and to health. It's an emergent quality—that sense of miraculous specialness or Presence we can't define, but we can feel. It's there. And we recognize containers, vessels, places and times that are meant to protect and garb that holiness.

A sanctuary usually has some border, a way to let someone know that they are entering "holy space." But what makes one place more holy than any other? Isn't the whole idea behind religion that of Oneness? Can't God be found everywhere?

Of course, the answer is yes. God is everywhere, and yet we need sanctuaries. This is a core paradox: Judaism believes in Oneness, in the ultimate unity of all the cosmos—yet, Judaism also talks about separations: dividing between Jews and Gentiles, between Shabbat and the rest of the week, between the Holy Land and other lands. How can we have it both ways?

The same question was asked centuries ago when the rabbis asked how God could fit into the small space of the Holy of Holies, the central sanctuary of the ancient Temple. In the midrash, the rabbis struggle with God's Infiniteness, in contrast to human attempts to represent God as dwelling in a particular place:

> "Make for me a Sanctuary, and I will dwell among them." [Exodus 25:8]. At that time (King Solomon)

117

asked God "will God really dwell on earth? Even the heavens to their uttermost reaches cannot contain You, how much less this House that I have built!" (1 Kings 8:2) . . . . But God said, "Not according to my strength do I ask, rather according to your strength." If I were to request it of the entire world, it wouldn't be able to contain my Glory . . . rather, I only request from you 20 cubits to the South, 20 cubits to the North, and 60 cubits to the West." (Bamidbar Rabba 12:3)

It may be that God is Infinite, filling the entire world (and even then, not able to be contained), but for our eyes and senses, we need to deal with time and space, and that is ok with God. God desires to reside in the world "according to your strength" as the midrash puts it. We aren't gods who can always see through God's eyes. It's ok that we find God in this world according to our limited, human capacities. So, we have sanctuaries, vessels or containers of time or space where we make it a little easier to sense God's Presence. Training wheels. That's the way the world works.

But there are many sanctuaries. Some large and some small. Nested to fit the need of the hour. In fact, that verse quoted in our midrash, "build for Me a Sanctuary, and I will dwell among them" has been interpreted by the rabbis to mean not just "dwell among them" referring to the Sanctuary itself, whether that was the Tabernacle tent in the desert or the Temple in Jerusalem, but rather, "dwell in them" meaning inside us as people. We, in our society together, can be the real sanctuary; we, as individuals, in our bodies, are also God's sanctuary.

The ancient rabbis imagined the world in concentric circles of holiness, starting from the Holy of Holies. In the Mishnah (Kelim 1:6-9) the rabbis list ten levels of sanctity, being very specific about the laws and restrictions that apply to each level: the land of Israel, walled cities within the land of Israel, within the walls of Jerusalem, the Temple Mount, the Vestibule, the Women's Court, the Israelite Court, the Priest's Court, between the Hall and Altar, the Holy Palace, and the Holy of Holies.

In ultimate reality, God is infinite, filling all the universe; all is essentially One in holiness. But we can't always be aware of that ultimate reality. From our limited perspective, we need to work with divisions and separations. Apparently, we are not the only ones. The whole natural world works that way as well. An organelle such as the mitochondria is within a cell, the cell is within the organ, and so on. A tree in a forest, a forest in a bio-region, a bio-region within a larger ecosystem, etc. As we saw in the chapters on emergence, life itself began with a membrane, a vessel.

We don't need to decide between having boundaries and openness or between tribalism and universalism when we think of the way natural systems always are nested, with one entity fitting into another one, one whole which is a part of another—when we understand that in nature boundaries are rarely hermetically sealed. Membranes are essential for life, but they are permeable.[49] When we experience the patterns formed by all those parts together, we get intimations of their ultimate Unity. That, in fact, seems to be the way that we have been instructed, in the Torah and tradition, to remember God's Oneness: this secret of nestedness is hiding in plain sight in the central statement of Judaism: the Sh'ma.

Listen—Sh'ma, Oneness, and the Nestedness of Daily Life

The most well-known of Jewish sacred declarations, which my childhood Reform rabbi used to announce every service as "The watchword of our faith" is "Hear, O'Israel, The Lord our

---

49 The phrase "permeable membranes" was a favorite of Rabbi Zalman Schatcher-Shalomi. He used this biological image most often to describe the boundaries of the Jewish people—not completely open and not completely closed either.

God, the Lord is One." Or, in more contemporary language: "Listen, God Wrestler, YHWH is our God, YHWH is One." This is a quote from the biblical book of Deuteronomy, where it is followed by the paragraph which we also have adopted as what is sometimes known as "the V'ahavta"—"and you shall love:"

> You shall love the Lord your God with all your heart and with all your soul and with all your might. Put these words which I command you this day on your heart; teach them to your children, and speak of them when you sit in your house and when you walk on the way; when you lie down and when you get up. Bind them for a sign on your hand and let them serve a symbol on your forehead; write them on the doorposts of your house and on your gates. (Deuteronomy 6:4-9)

Traditionally, there is a phrase that is inserted between the biblical verses, right after the first line: *"barukh shem kavod malchuto, l'olam va'ed*—blessed is the majesty of God's glory for ever and ever"—which is said in a whisper. This set of statements, which we perhaps say so often that we become dulled to their power, are a meditation on the apparent paradox of nestedness and Oneness. It is the secret path, hidden in plain sight, to a different way of feeling ourselves in the world—a way which can help us break free of the either/or way of thinking that plagues us today.

How should one absorb the central idea of Oneness, or the Unity of All Being, into one's consciousness? The answer is that one must place reminders at a number of transition points between the nested domains that structure our experience. Put these words at the liminal points between your inner world and your outer world, first starting with your own bodies—place it on your heart. But don't stop there. Put it on your arm as a reminder to integrate it into your actions, and on your forehead to remind you to use your eyes to see the Divine Unity wherever you look. Speak of it in your house and when you walk on the road, in your communities and public forums. Place it on the doorposts of your house and also on your gates, marking two transition points between home and outside. And teach your children, so that this message stretches over the borders between generations as well.

At those places of change, of leaving one realm and entering a new one, be it the realm of my inner thoughts to an awareness of my physical body; or going from my house to my yard, then to the outside world; looking from my generation to the one growing up after me—we are enjoined to recommit to these words. Paradoxically, in order to come to know Oneness, we need to pay attention to all our points of division.

Fractals of Activism

It may seem obvious when we are talking about a religious practice such as reciting the Sh'ma that we should start with our inner lives. Religion, after all, is empty if it is not built on sincere belief and deep feeling. But it is perhaps less obvious that we similarly need to start with the heart and soul and build out when we approach our many challenges in the social, environmental and political worlds. There is a temptation to just get out there and act. That's activism.

We are often caught up in the wonky, techie, details of how we are going to get out of whatever mess we find ourselves in. We often try to argue the clear rational points that will make others see it our way, or work to elect candidates who will turn the country around; and those are all good things to do. But we sometimes forget to heal ourselves. We forget that we are nested within concentric circles and all of them count.

adrienne maree brown is an activist who has written a wonderful book called *Emergent Strategies: Shaping Change, Changing Worlds*. She writes to her fellow young activists about ideas very similar to those in this book, but in a more fluid and informal register. What I'm calling nestedness, she calls fractals. That works, too.

In any movement work, she says, you have to walk the walk, make it real in your own life, not just push your views out to others. It is about how you are with your family, friends, and community. "Fractals" stands for how we exist on various levels, and we need to take care of all of them. It is a creative and fluid way of telling our story of connectedness while maintaining our sense of uniqueness. She quotes this poster she saw on the wall of one of her mentors, Grace Lee Boggs. It said, "Building Community is to

The Collective as Spiritual Practice is to the Individual."

The analogy: "building community is to the collective as spiritual practice is to the individual"—challenges activists to work on building our individual core, our values and our perspective, through spiritual practice as much as we work on building community. Our spiritual practice can "pull us together," allowing us to hold the many parts of ourselves without breaking down from inner conflict, seeing ourselves as part of a larger picture. This is as necessary to activist's work as building community is on the collective level.

brown offers practical ways to assess how one is doing in aligning one's individual values to the group, so that there is resonance across fractals, the different levels of organization. For example, she offers this worksheet:

Assessment of Fractal

o   Are you a perfect living realization of your values and beliefs?

o   Is your group a perfect living realization of your collective values and beliefs?

o   What are you embodying in your daily life? In your work?

o   Individual: Interview three people you trust in your extended community to give you feedback about how you show up in the world. Share your purpose/ intention with each of them and ask them to hold that as they answer your questions. Sample questions:

> o   What is my impact in the world?

> o   In three words, what am I embodying?

> o   What do you think I could grow?

o   Organizational:

> o   Interview three people in the community your group/organization serves to give you feedback

based on how y'all show up in the world. Share with them what you think you are embodying and have a brief discussion on how much you are or are not embodying that.

o   Can everyone in the organization state the vision and mission accurately, even passionately?[50]

That fractal point of view: recognizing and giving our attention to the many levels of organization in which we participate, goes for all the kinds of work we do. Here's an important example: climate.

Our Fractal Challenge—The Climate Crisis

When we are dealing with our climate crisis, we can't stop at the borders of our country, but must see ourselves as nested within the larger whole of the planet earth. This is one of the fundamental shifts that Rabbi Zalman Schachter-Shalomi constantly spoke and wrote about, reminding us of the iconic photos of the earth from outer space which appeared in the '70s as something that pulls us into a new, Gaian consciousness. But there is always going to be a temptation to "beat" the competition, to have our country come out on top in the race for economic and political power. Gaian consciousness or not, we're unlikely to give up our national borders or the sense of competition among nations anytime soon. These rivalries are real in our world.

So, how do we cooperate with our competition? It's not so different from any sports team. When I played basketball on my high school team of course I wanted to be a star (I was far from being one). I wanted to score as many points and make as many outstanding plays as possible. I was, we all were, in a competition to be the stand-out. But that didn't stop us from also working together, from sometimes taking one for the team, from passing the ball when I wanted to shoot, to setting a screen so my

50 adrienne maree brown, *Emergent Strategies,* (Chico, CA, AK Press), 2017, 185.

teammate could score—moderating our competitive instincts in favor of realizing that we'd all succeed more if we succeed as a team.

We can't hold on to the utopia of no countries, no national interests, all one humanity. That may be the messianic vision, but Judaism has generally held the principle that we need to work with the world as it is—and move it in stages toward the vision. To achieve the goal of unity, we need to work with the divisions and the fractal reality of life.

The 2015 Paris Climate accords were a powerful step in the right direction, as representatives from nearly all the world's nations hammered out a balance between their own interests and those of the whole. It wasn't enough, it was not the accord that it might have been, but it was a movement in the right direction. They came together and could move forward because they held that tension and realized that there was a need to work together without giving up on their self-interests. We need to see ourselves as parts of the larger whole of the planet when we are dealing with crises that operate on a planetary scale. We need a both/and attitude, not an either/or.

Understanding our nestedness goes for nations vis-à-vis the planet, but it also applies all the way down the scale to the local and personal levels as well. As Bill McKibben has written in his powerful book *Eaarth: Surviving on a Tough New Planet*, we must now adjust to living more locally as the supply chains and energy needs of the global economy become less viable. He points out that this will make us happier and safer, as we strengthen the local relationships that give us the feeling of community and belonging to a place; and as we rely more on local institutions which are less likely to fall prey to global, international trends. As a whole, the redundancy of many local solutions works the way it does in an ecosystem, making the whole larger system more resilient. Many nested communities make for a more resilient way of life than reliance on a few global corporations or conglomerates.

Wendell Berry wrote back in the 1970s about "cheap energy mind"—the way that we have grown used to relying on distant, paid-for services and specialists instead of our own hands, our families and our neighbors for life's needs. Berry writes about how we are the first generation of humans who don't feel we need our neighbors, and so we often don't even know them. Yet, there is something that feels intuitively wrong with this picture. As much as we feel the urgency to work on a big, national and international scale, our small, personal and interpersonal decisions are also part of the picture.

When Nadav was still a small toddler, we moved into our present home in Newton, Massachusetts. We were hoping that we'd make some friends and connections with our neighbors. We'd already learned the truth of the saying "it takes a village…" when it comes to raising young children. The idea that child-rearing can be done in an isolated, nuclear family is simply crazy. It's not the way humans evolved, and it's not the way humans have parented for almost all of human history. Our direct neighbors on either side were very welcoming, and we've become good friends and allies to one another in many ways. But maybe because this is New England with its famous reticence and reserve, especially regarding new-comers, or maybe because of that cheap energy mind that Berry was writing about, none of the other neighbors on the street so much as acknowledged our arrival.

With a little Jewish chutzpah, we decided that wasn't something we were going to leave up to them. So, we put notes in everyone's mailboxes on our street that we were having an open-house on the next Sunday afternoon, and they were all welcome. We put out homemade pumpkin muffins and hot apple cider of course, but we also put up a big poster board with spaces where people could fill in their recommendations for the best car mechanics, dentists, hair salons, kid-friendly places to eat, etc. People came, and with that little nudge, they were mostly friendly and helpful. We now have five families on our block where we can call on each other to borrow a tool, or trade veggies from our gardens or give one another drives to the airport to avoid big taxi or Uber fees.

Sometimes it's just about borrowing a cup of flour or a bit of company, and sometimes it's a little higher-stakes. Once, when I was out of town for work, Ilana caught a bad flu and lost her voice. Nadav got sent off with some school friends and our grandfatherly neighbor down the block came over and read seven-year-old Eiden's favorite book of Rick Riordan's Greek myths for about two hours until Eiden fell asleep (and so did the convalescing Ilana on the couch).

One of our next-door neighbors is a veterinarian, and when my eighteen-year-old cat, Lulav, came to the end of his life, she provided the injection to euthanize him, and sat with us in our bedroom while I held Lulav and Eiden cried (and so did Nadav, Facetiming in from summer camp). Those same next-door neighbors just last summer organized with us and another neighbor to buy a top-of-the-line, ultra-safe trampoline and, since their son Mack is very good buddies with Eiden, they decided it made sense to put the trampoline in our yard. They aren't Jewish, but Mack has become a big fan of my challah and Ilana's matzah ball soup.

The neighbors on the other side have been life-savers in many ways. For one example of many: when Nadav was tiny and getting up at 5am every morning, and I'd be away for work, Ilana would wait at the window looking to see when their light would come on. Then she'd go and quietly knock on their door and join them in their kitchen, everyone in pajamas and bathrobes, for a cup of coffee while their 12-year-old twin girls played games with Nadav on the rug. Not only have our lives been greatly enriched by our connections with neighbors, but it's hard to imagine how we could've made it this far without them. Ilana and I smile remembering these and many more stories that have become part of our family's lore and the fabric of our kids' youth.

Back to climate change: Bill McKibben writes in his latest book, *Falter*, about just how radically the discovery of fossil fuels changed human life. He writes about how human history has mostly been a story of expansion, but the expansion brought about by fossil fuels outstripped everything that had come before:

> Part of the expansion was literal: instead of being confined to the few villages where a horse or your feet could carry you, everyone was able to move about, a liberation from geography that changed everything, right down to who you might marry. And cheap power led, at the turn of the century, to air-conditioning, which in turn meant that places once so hot as to be marginal were now "the Sun Belt." But the biggest part of this expansion was economic: everyone in the Western world now had access to, in essence, slaves who would do an absurd amount of manual work. A barrel of oil, currently about sixty dollars, provides energy equivalent to about twenty-three thousand hours of manual labor.[51]

Fossil fuels have allowed us for about the last one hundred and fifty years to live incredibly expansive, convenient lives. But all that expansion has also eroded our sense of what it means to live in an intimate, nested community. We have learned to live with an unbelievable amount of independence, and also loneliness. Loneliness brings on a culture of addiction—we crave something—connection, belonging—but we are taught, seduced, marketed, into looking for it in all the wrong places. The easy availability of convenience brings on an obsession with never losing those conveniences, with hyper-control over our environment. As we saw with the river valley cultures of the ancient Near East, we have become obsessed with eliminating all contingencies, reaching for complete control. This too is another addiction.

---

51 *The Sun Magazine,* October 2019, "A Shrinking World," 14. Excerpted with permission from Bill McKibben, *Falter: Has the Human Game Begun to Play Itself Out,"* (New York, Henry Holt and Company), 2019.

All of this expansion and convenience is alluring and isn't making us happier. We have everything and nothing. We have the wide world at our fingertips, but no sense of community or belonging or purpose. That would be bad enough, except that now we are realizing that the very source of our ability to live those expansive modern lives, the fossil fuels which give birth to "cheap energy mind" is itself killing us. The revolution of fossil fuels has launched us to new heights, and it is at the bottom of not only our social psychological crises, but our climate crisis as well.

So, it does make a difference that we change our habits and our life-style, even if it seems like we're not making enough of a difference. As adrienne maree brown wrote, we need to walk the walk and talk the talk. As Bill McKibben says, we must re-learn the pleasures of smaller, local economies and communities. We need to mirror nature and change ourselves, starting on the smallest fractal levels, or we will never change the big industrial and national players. Management consultant and writer Peter Drucker made the famous statement, "culture eats strategy for breakfast."[52] Meaning that we can't just strategize our way out of this problem, we need to change the culture, and that means changing ourselves and the many small actions that make up our collective culture.

I know for me, riding a bike gives me satisfaction and the endorphins that push the pleasure points in my brain. I just feel good after working up a sweat riding to work or meeting friends for a weekend softball game. And I've noticed little things about our neighborhood that I wouldn't have if I sped by in a car. This house has an especially nice garden; that one just put up a sign supporting a candidate in the local election, someone is cutting down a tree or re-painting their home. I'm a bit more intimately connected to the neighborhood. It's one little action that has

---

52 This seems to be folk knowledge, which apparently started to make the rounds in 2006. There is general agreement that Peter Drucker did say this, although an exact source didn't come up in my search.

changed my culture just a little towards connection, real feelings of pleasure in moving my body, and away from the addiction culture of fossil fuels.

Perhaps when we all start to change even little things in our culture, the positive feedback can shift things more quickly than we had thought possible. Maybe it comes back to that old bumper sticker: "Think Globally, Act Locally" which captures the nested reality of how life works. We want and need to see the big picture, but if we really want to succeed, we need to work toward our solutions locally as well, in smaller units.

This paradox of how the interweaving of the macro and the micro leads me back to the Sh'ma and the ways that I see traditional Jewish understanding holding on to the paradox of separation within Oneness. But this embracing of division and separation, even in the ultimate service of Oneness, brings up one of the thorny questions that Judaism has faced in the modern world—why do we separate ourselves from others? Do we think we're better? What is all this about being the "Chosen People?"

The Paradox of Choosing, Separation, and Oneness

When we look at that statement of God's unity, "Listen God-Wrestler, YHWH is our God, YHWH is One," we might notice a complication: It's not simply saying God is One. It's saying two things—YHWH is Our God (*eloheinu*) and YHWH is One. What do we mean when we say "our" God? What have Jews meant when we've claimed, for hundreds of years, to be God's Chosen People?

First, a little historical background: according to historian Reuven Firestone, in the ancient Near East all the peoples were chosen by their gods.[53] The basic understanding was that each god took care of their particular people, who were living on their particular land. When a person travelled to another land, one naturally did the respectful and reasonable thing by offering sacrifices to the god of that land. After all, my god was back home and was not in power around here, anyway. I'm chosen by my god, but I respect the fact that you're chosen by your god as well. Polytheism could be quite tolerant.

When monotheism emerged and the God of Israel evolved into being understood as The God of the world, we found ourselves in a kind of awkward hybrid situation. We kept the ancient idea of a god protecting that god's own nation or tribe, but now all of a sudden, our particular god is also the God of the world. This has come out badly in many ways.

In various periods and places, having the One God of the World on one's side has invited the idea of converting, sometimes forcibly, other people to this One True God. That has been more common in historical memory in Christianity and Islam, but it is

---

53 Reuven Firestone, *Who Are the Real Chosen People? Chosenness in Judaism, Islam and Christianity* (Woodstock, VT, SkyLight Paths Publishing) 2008.

not unknown in Judaism in earlier times. More often, especially in the Diaspora, Judaism has been in a position of keeping to itself, without much power to convert others. In pre-modern times, when it was the norm for peoples, ethnic groups, to stick to their own, that was seen as generally ok, even if sometimes people resented Judaism's claim to being the Chosen People. Sometimes people even admired the Jews for minding their own business. But in the modern world with its values of universalism, this has become more problematic.

Judaism has been accused of a kind of superiority complex and of disdaining other people. When, in the period of European emancipation, European nations were considering offering citizenship to the Jews in their midst, they would often note that the Jews themselves wouldn't agree to fully join in the life of a citizen. They would continue to stick to their own, marry only one another, disdaining the wonderful bounty of equality and freedom being offered.

Most Jews did, in fact, accept with eagerness the opportunity to expand their lives and enjoy the economic and cultural richness of the wider society. But Jews didn't leave behind the idea of chosenness. Sometimes this idea has actually gone along with feelings of superiority to others. This has been exacerbated, of course, by the fact that Jews in many times and places have been an oppressed minority and oppressed minorities often think of themselves as chosen; as smarter, more resourceful, and morally superior to the ruling majority. Perhaps not so much as an ideology as a coping mechanism.

But, clearly, today many people, me included, understand that the idea of chosenness is problematic. Jews are not better than other people. We can see that there are many cultures, religions and paths toward a good, moral life. There are many paths toward holiness and sanctity. Who are we to say we are chosen? Perhaps we should just get rid of the idea and adopt a purely universal God.

Some Jews, in the Reconstructionist Movement for example, have done just that. But I don't think we should throw out chosenness so quickly. My answer is that we must come back full circle to the roots of what chosenness originally meant: a special, personal relationship between a god and a particular people.

While we don't need to keep the ancient understanding of many gods, one for each nation, we can find a subtler understanding of what one God means—one God with many faces; one God with many names; one God with many paths to reach Him, Her or It. But, coming back to the ancient idea of a personal relationship with our God seems to me a good idea. We all deserve to feel special.

I use the analogy of the family to illustrate what I mean. My kids happen to be the most brilliant, beautiful and talented children in the world. That is how I feel. I would guess that, if you have children, you might feel the same way about them, too. And that is as it should be. Every child should feel that they are loved like no one else; that you are not just their parent by chance, but that you would choose them out of every other child in the universe to be their parents.

That kind of love is like fertile soil for a soul to grow in. Of course, parents also need to discipline and set limits for their children, know when to tell them they're not the center of the world, and all the rest, but that special love that says, "you are special to me"—that is a wonderful thing for any child to have. The thing is: I have to remember that this is only my subjective feeling. My kids aren't really better than other kids. Other people's children are also special, and their parents think they're the greatest as well. It's the same thing for religion.

I understand their reasoning, but mostly disagree with the Reconstructionist liturgists who take out any references to chosenness in the prayer book. I think it's an important part of Judaism that we have always felt a special relationship to YHWH, HaShem, to the Rebono Shel Olam (Master of the Universe), to Gottenu (literally, "my little God"), and all the many personal names Jews have used throughout the centuries. That is as it should be.

The *Pearl* and the *Flame*

Abraham Joshua Heschel, in his classic work The Prophets, makes it clear that the biblical prophets were not simply reporting news and insights from a distant, abstract philosopher's god, but were intermediaries communicating the divine pathos, the caring of an involved, deeply caring God. That is something worth keeping.

But we need to remember that feeling chosen as a religion, like our feelings for our children, is our subjective experience. Other people can be chosen as well. I heard it quoted that when the Dalai Lama met with the Jewish delegation that visited him in Dharamsala he said, "I love that the Jews are the Chosen People! We are also chosen!"[54]

So, as has become the practice in many Jewish Renewal communities, when I come up to bless the Torah in synagogue, I don't say *asher bachar banu mi kol ha'amim*—"[blessed is the One] who has chosen us from all the people of the world." I change one letter in the Hebrew and say *asher bachar banu im kol ha'amim* —"[blessed is the One] who chose us *along with* all the nations." We feel chosen, and we acknowledge that other nations feel chosen as well. That is well and good, you may say, but we still have the problem of difference—how can we Jews claim that all is unified in Oneness, but still have so many boundaries and make such a point about separation?

Havdalah—Boundaries and Beyond

When I was in my early 40s I had a serious girlfriend who wasn't Jewish. In fact, she was (still is) ordained as a liberal Christian minister, and some people call her by her Sufi name, because she had been quite involved in that community as well. How did I, as a rabbi ordained in the Orthodox tradition, come to

---

54 This was most likely Rabbi Zalman Schachter-Shalomi, *z"l*, though, my memory isn't clear on this. In any case, it seems that the Dalai Lama made this statement on more than one occasion. Rabbi David Wolpe reports nearly the same words in his conversation with the Dalai Lama as reported in "A Holy Man of Laughter: My Encounter with the Dalai Lama" in *The New York Jewish Week*, June 12, 2018.

be going out with her? To make a long story a bit shorter: In my typically perverse way of doing the opposite of the usual "seeker" path, instead of rejecting my religion of origin and going as a young adult to India to seek true spirituality, I deeply embraced my religion of origin in my 20s and 30s, and, when everyone else was settling down with jobs and family, I ventured in my early 40s to India and became involved in meditation, Eastern spiritual paths and New Age philosophies.

I met my Sandy (not her real name) in Dharamsala, India where she was on a Buddhist meditation retreat and I was helping to lead a Passover Seder for the many young Israeli seekers who were in Dharamsala, along with various Tibetan Buddhists, local Hindus, and travelers of all varieties. We, all 150 of us, made a wonderful, do-it-yourself Seder, sitting on pillows on the floor of a half-built hotel, and drinking the four cups of "wine" which we had made by soaking raisins in big plastic bags of water for a couple days.

Sandy came to the Seder as well as to the various hikes, discussions, sing-alongs and just hanging out that we did all Passover week. We felt an immediate connection, and when we both returned to the New York area, we jumped into our interfaith adventure. She found many aspects of Judaism attractive and I was being opened up to a whole world of New Age spiritual paths and practices. We were feeling our way forward to see where this would lead. But, in the end, it didn't work for us. In our case, the differences in our belief systems were too far apart. One person I spoke to for counsel during this time told me metaphorically: It's okay if you're in different cars of the train, but if you're riding different trains, that's not a good sign.

I remember one Jewish practice that perhaps came to stand for a key to our different ways of looking at the world. We had a lot of trouble with *Havdalah*. *Havdalah* is a short ritual involving blessings on wine, a braided candle and spices, performed on Saturday night to mark a separation between Shabbat and the rest of the week. Many people find it to be one of their favorite Jewish rituals. It's short, musical, and stimulates all five of our senses: taste, smell, touch, sight and sound. But Sandy hated it.

She couldn't understand why we wanted to say that holiness ended here. Why did we want to put down the rest of the week? Wasn't it obvious that one could find holiness on any day of the week?! What she was asking was: Why create separation instead of unity? Jews get asked this a lot—why are you staying separate from people? Lurking right behind this is the question: Do you think you're better!?

The Saturday night Havdalah ritual, instead of being the inspiring beginning to a new week that I've often experienced, with Sandy more often led to an argument, hurt feelings and not much enlightenment. It felt a bit like we were speaking different languages. What I tried to tell her (I probably wasn't at my most eloquent or charming in these moments) was something like: No. I don't think Jews are better than other people, and I do think that there is holiness everywhere, any time. It's a paradox of separation and unity that I don't think I was ever able to fully explain in a way she could hear. Saying havdala and marking that separation just didn't imply for me what it did for her.

Permeable Membranes

The whole concept of separation has often been misunderstood. As we discussed in the last section on emergence, life is only possible with a membrane. But the membranes aren't meant to be hermetically sealed. They make it possible to have an entity, which then connects to others. The way of nature is not an undifferentiated unity, but an interconnected, nested pattern of units.[55]

We live in a dynamic tension: on the one hand, we understand that, in the deepest, ultimate truth, there is holiness everywhere. In other words, every day is Shabbat. The World to Come is already here, if we would only open

---

55 Rabbi Zalman Schachter-Shalomi would often use the phrase "permeable membranes" when talking about the various separations in Judaism, and, especially between the Jewish people and other peoples.

up our eyes! But, on the other hand, we understand that if we tried to live that way all the time, we'd probably fail.

Unless we were the highest tzaddik (saint), the most enlightened being, we'd be worn down by the world, which is not operating on that level. If we tried to get our work done, pay the bills, argue with the bank, listen to the news, get our kids to school on time, and still be on the level of Shabbat, constantly aware of the holiness all around us, that's a great goal, but most of us would end up losing our sense of what we're talking about. We'd muddy up the waters so much that the holiness would get lost.

This tension is why we have boundaries. Shabbat is supposed to be a guarded, sacred time of holiness. It is a time we can slow down, leave the striving, the commerce, the struggles and the distractions out, and really allow ourselves to experience the world as sacred. So, we put boundaries around it. We "guard" it with rules which separate it from the rest of the week like not using electronics, or not spending money. We make it special by eating particularly yummy food, singing during meals and talking about spiritual ideas instead of work-related matters.

These things make for a sense of specialness and separation: holiness. But, ultimately, that sense of the holy is supposed to seep out into the week. I've known mystically inclined Jews who would have the practice of greeting people on a random Wednesday or some time in November nowhere near a Jewish holiday with *"Gut Shabbes, Gut Yontif"* (Good Sabbath, Happy Holiday!) as a way to remind themselves and us that the holiness of Shabbat and Holidays are really always here, beneath the surface, waiting for us to notice them, in all times and places.

We may not observe Shabbat the whole week, but we try to punctuate our days and weeks with reminders so that we carry Shabbat holiness with us as much as possible. Traditional Jews stop our work lives every day to say prayers; we stop before and after eating a meal to say blessings. We take little rests in our week to remind us of the same thing that Shabbat reminds us of: God is the Source, all this is a gift, be grateful. Don't take it for granted.

And, even though we guard the sanctity of Shabbat, the week

seeps into it as well. We traditionally don't handle money on Shabbat, but we might pledge *tzedakah*, charity, in the synagogue service. We don't traditionally practice medicine and healing (except in emergencies), but we make a special point of praying for those who are ill. We traditionally don't cook food, but we find ways to heat things up, to prepare an enjoyable meal. These aren't desecrations of Shabbat. They are ways that the lines are a bit porous, enough to keep both Shabbat and the week distinct but in relationship, and to keep both vital and healthy.

It is a two-way flow—the Shabbat needs a bit of the week and the week needs a bit of Shabbat. The goal may be to experience every day as just as holy as Shabbat, but the way that we get there is not to mix everything up all at once, but rather to create a pattern. There is a back-and-forth rhythm of alternation between work and rest that gets the energy flowing between Shabbat and the week and draws out the best in both.

Rabbi Arthur Green likes to note that this dynamic works in the three primary realms or dimensions of space, time and person (*olam, shana, nefesh*) as defined in the *Sefer Yetzira* (and also often quoted in one of his favorite Hasidic books, the *Sfat Emet*). Just as there is a special, holy time, Shabbat, there is a parallel special, holy people, the Jewish people, and a special, holy place, the ancient Temple, or sometimes the Land of Israel. In all three, the intention is not to separate permanently, or make a statement that these are essentially better than the rest. Rather, in each realm, the intention is for these to be examples, training wheels, for the time when their holiness will be recognized in all people, all lands, all times.

It is like music: without the separations between tones, between sound and silence; the energy and tension that is created by harmonies and the movement of rhythm, the music wouldn't be music. Listening to a song, we might love one note, and wish that the singer would just stay there forever, but in reality, the beauty of that note is only revealed in the pattern of all the other notes, tones and rhythms. Miraculously, from those patterns of sound, silence, harmonies and rhythms, we get an intimation of oneness that emerges from the differences. If we allowed everything to flow together, we'd actually lose the beauty.

I find that same intimation of Oneness in nature. When I'm walking in a forest, beach or field, there is something in the patterns that sings that Oneness. The biblical Psalmists knew this music well as we see, for example, in Psalm 98:

> Let the sea and all within it thunder,
>> the world and its inhabitants;
> let the rivers clap their hands,
>> the mountains sing joyously together
> at the presence of YHWH,
>> for God is coming to rule the earth;
> God will rule the world justly,
>> and its peoples with equity.

In this and other psalms, the world of nature appears to be out in front of us humans—they already sing the song of Oneness in all their diversity. They sing in anticipation of a time when we will catch on; when all the diversity of humanity will harmonize and sing that song of Oneness. It's a lesson that we're in dire need of learning. The mind-set of nestedness, of Mikdash, which allows us to hold that paradox of unity within diversity, creating a "both/and" instead of an "either/or" reality, is deeply woven into Jewish tradition. One of the most beautiful strands of that tradition is Kabbalah, where nestedness is key to understanding how the world, with all its mixed blessings, can also be experienced as a sacred Unity.

Kabbalah—A Language of Holiness Hidden

"The external shells of the higher worlds – they are the inner cores of the worlds below them."

— Kolonymus Epstein of Krakow, *Ma'or Ve'Shemish*, B'Shalach

Kabbalah, the classic form of Jewish mysticism, teaches that holiness is everywhere, but that it is hidden. It is one of the most important languages or systems of thought that Judaism has developed for expressing the organic, living patterns of our world. Kabbalists love to use metaphors such as sparks, garments, husks, and shells. There is a lot of interesting theory and lore behind these metaphors, but they are all essentially saying that everything in the universe is a part of God. God is the essential energy and life of the cosmos, yet that holiness, or Godliness, of necessity, is covered or hidden.

Using the analogy of fire or light: since God's light is infinitely bright, God's fire is infinitely hotter than a million suns—it cannot be directly revealed, or else it would burn, consume and envelop everything in its brightness and heat. What we know as Creation, or the Cosmos, would simply revert to Endless Divine Oneness. So, in order to have a world, a world of separate existence, of stones, trees, people, animals, stars, sunsets—all time and space— God hides the divine light and only emanates that light or energy into the world in stages, or degrees, covered by garments.

Sixteenth century Kabbalist Moses Cordovero put it this way,

> When powerful light is concealed and clothed in a garment, it is revealed. Though concealed, the light is actually revealed, for were it not concealed, it could not be revealed. This is like wishing to gaze at the dazzling sun. Its dazzle conceals it, for you cannot look at its overwhelming brilliance. Yet, when you conceal it— looking at it through screens—you can see and not be harmed. So it is with emanation: by concealing and clothing itself, it reveals itself.[56]

The paradox that the light of God must come into the world in garments, must be covered to be seen, is at the core of a kabbalistic world view.

---

56 Moses Cordovero, *Pardes Rimonim* 5:4, 25d. Quoted in Daniel C. Matt, *The Essential Kabbalah* (New York, Harper Collins), 1996, 91.

We can use our own experience of creativity to get an idea of what this means. When an idea pops into my mind I often wonder where it came from. It's somewhat of a mystery. I might say that the idea was floating around the ether, completely unformed. Then it takes on a "garment" of a thought in my head. That thought might not even have any words at first. But the next thing that I would do, before that spark of insight disappears, is to put it into words, at least in my mind. That process might be so quick that I don't even notice it. So, now this spark has taken on the garment of words. But as any person with a creative idea knows, it is all too easy to lose it if you count on just keeping it in your mind. I'd probably want to go write it down. That's another layer of garment.

We can pause here to notice something about this process. It can sometimes be difficult to translate that thought in my mind into words on paper. Often, I feel as if I've lost something in the translation. It's hard to capture that original spark, but I must do it if I want the idea to have any real existence in the world. The garment is essential, but it can come at a price.

Very often, depending on what kind of creating we're talking about, there are more steps. If I'm a builder and my idea was to build a house, then I have to think about how I'm going to translate from my blueprint on paper into a real brick, wood or concrete house. Again, I'm going to perhaps lose something in the translation. Anyone can enjoy my drawings and dream about my amazing house. I can put it on the internet and millions of people can all marvel at the creative idea in the blueprint. But I definitely also gain in making it someplace a person could actually live, rather than just dream about as they look at the plans. Garments are a mixed blessing. They are necessary to make a world manifest, but they can also obscure the original spark. They create individual things in place and time, which also creates separation, distinctions, and divisions.

Kabbalah looks at the whole world as coming into being in that same way that your idea comes into being. The basic kabbalistic language is of the emanation of the ten Sefirot. Everything starts with *Eyn Sof:* ultimate, infinite Endlessness. This unknowable Source of All, emanates its light into vessels, each interacting with and balancing one another in pathways and patterns, personalities and qualities, until a world manifests.

Another one of the ways the Kabbalists conceptualize this series of increasingly concrete or material "garments" that take us from an insubstantial spark to an actual, physical object is to posit four "worlds" that take us from the mystery of God's essence to the world that we know. The highest world is called *atzilut*—emanation—the nearly pure spirituality that The Endless One sends into existence. Next comes *bri'ah*—creation—this is often associated with intellect and understanding; after that comes *yetzirah*—formation, associated with emotion and creativity; then finally *assiyah*—the physical manifestation. All of reality is a nested series of "worlds," increasingly concrete "garments" of the infinite light of God.

Sometimes the garments can be too thick. That's when we start calling them "shells" (*kelipot*). When we start to see too much blockage of the light; so much separation so that we don't see any unity; so much difference so that there is no harmony—that is no longer a useful garment but a stubborn shell.

For example, say I'm walking down the street and I'm attacked by a mugger. I've got a couple of options. I can run or fight or I can appeal to the mugger's better nature. In kabbalistic terms I might say that there is a thick shell covering this guy's divine light. If I am a highly developed *tsaddik* (saint) and I have confidence that I can shine my light and arouse his inner light by shining my love onto him and drawing it out, that's great.

But I may not be so confident that I can do that. In that case I have to take his threats of violence against me very seriously. His threats to take my money or beat me up may not be the ultimate truth. They may be only the result of a "shell" covering his true, inner divine light. But as long as I'm living in my shell (my body), I can be really hurt by that shell. So, in my case at least, I'd run, or give him the money, then run.

We can see from this small example that the idea of nestedness can make for some interesting options in dealing with reality. On one level, that guy is dangerous, bad. But only on one level—on a higher (or perhaps, deeper) level, he's pure light. It's not either/or. We need to work with the separations and divisions and garments in the world. Sometimes, as in this case, they are too thick, masking the light and letting evil manifest, but in many more cases, garments are a necessary ingredient in living.

Compost, Clothing and Kabbalah

If you're the one being mugged it doesn't feel trivial, but it's admittedly a pretty trivial example of the ways that a kabbalistic world view of sparks, garments and shells can shift our perspective on very consequential life changing issues. It changes our way of thinking about good and evil, suffering, forgiveness and renewal. When we understand that everything is ultimately a part of the One, then there can never be anything that is purely evil. There may be some pretty thick shells, there may be things and people that we need to treat as serious threats, people and things that we may need to fight, but, in the end, there has to be a spark of God in there somewhere. That makes a big difference. It means that nothing and no one is completely irredeemable. It means that ultimately there is no such thing as waste. Only covered sparks. And one of the best ways I know for learning that lesson on a daily basis is by composting.

I've been composting my kitchen scraps for at least a couple of decades and I consider it one of my core spiritual practices. It teaches me day in and day out the lesson of hidden sparks: about the ways that what we thought was the smelly "waste" of banana peels and onion skins actually contains the potential to become fertile soil, and ultimately come back to life as new fruits and vegetables. It teaches me that *tshuvah* (returning to our true selves after losing our way) is real and embedded in the nature of the living world. I know that composting has changed me because it pains me when I'm in some circumstance where I'm forced to put my kitchen scraps into garbage headed for the landfill. It feels like a small desecration.

Covering and containers are key to composting. It's a process of transformation that is natural—those vegetable scraps will decay one way or another—but in order to effectively work with that process, it's very helpful to use some kind of container and cover. It makes the whole process more efficient and it protects us from the most unpleasant sides of composting. When people are starting to compost the main thing that sometimes trips them up is that they don't know that you need to cover the compost.

If I want those smelly kitchen scraps to have the chance to go through the decomposition process of breaking down and becoming soil in a way that works for all of us, I need to cover them with some hay, dry fallen leaves or grass clippings. If I don't, they are a foul, fly attracting nuisance. I've spoken to many people who tried composting but didn't practice the covering part—they usually give it up after a few weeks. It's like that Talmudic saying that I learned so many years ago from Ya'akov in Yerucham— miracles only happen when they're hidden from the eye. The miraculous transformation that happens in a compost pile needs a container.

David Abram, in his book *Becoming Animal* connects our own experience of the hidden, dark realm of sleep to the process of composting:

> We sleep, allowing gravity to hold us, allowing Earth—our larger Body—to recalibrate our neurons, composting the keen encounters of our waking hours (the tensions and terrors of our individual days), stirring them back, as dreams, into the sleeping substance of our muscles. We give ourselves over to the influence of the breathing earth. Sleep, we might say, is a habit born in our bodies as the earth comes

between our bodies and the sun. Sleep is the shadow of the earth as it falls across our awareness.[57]

We share with the vegetable world this same need for a container, a dark, hidden place to transform and renew. In general, we usually need some kind of container or vessel in order to bring out the sparks of holiness in our world. Some people have an ideology of not covering—they want everything to be out in the open and "natural." This has been a predominant theme especially since the 1960s and the counterculture. It has often been an important corrective to too much covering, too many restrictions, and it has brought out many important openings and liberations, but this ideology of pure openness has also had negative effects. Life actually works better with some coverings, some external vessels to hold us and give shape to our lives.

To take an example from a completely different realm, that of the practice of law, one of the intriguing facts that stayed with me when I read the 2008 book by behavioral economist Dan Ariely *Predictably Irrational: The Hidden Forces that Shape our Decisions,* was about lawyers and oaths. Ariely describes how the professions such as law and medicine had evolved from religious, esoteric knowledge which was held as a sacred trust: it was restricted to those who held the secret knowledge but these adepts would be bound by sacred oaths to use their knowledge for good.

These professional oaths lasted for centuries, but in the 1960s, with the movement to loosen up such stuffy, elitist traditions and deregulate, they let go of the requirement to take an oath on entering the profession. He describes how, since that time, the ethical behavior of lawyers has declined. Instead of individual freedom, flexibility and natural judgement flourishing, we've seen more greed and cheating among lawyers.[58] The seemingly ritualistic, purely external vessel of an oath apparently does make a difference.

57 David Abram, *Becoming Animal: An Earthly Cosmology* (New York, Vintage Books), 2010, 24.
58 Dan Ariely, *Predictibly Irrational: The Hidden Forces that Shape our Decisions* (New York, Harper Collins),2008, 209.

The *Pearl* and the *Flame*

Ana Levy-Lyons, in her book *No Other Gods*, deftly re-interprets the Ten Commandments as deeply relevant to contemporary progressive values. But in introducing her theme of the relevance of these ten ancient rules, she describes what she calls our "freedom fetish," our extreme resistance to any suggestions of rules that limit our freedom.

> Our love of freedom has become a fetish. The honoring of individual freedom over communal flourishing is a ubiquitous and powerful norm in this country among both progressives and conservatives, although in different ways. The trend in our culture has been inexorably toward a world of individuals, each doing his own thing. We elevate the self to an almost godlike status.[59]

Levy-Lyons pin-points the ways we viscerally resist any restraint on individual freedom. She describes a thought experiment that she did with a group of religious liberals. They were asked to imagine a tight-knit community that seriously followed their religious tradition. What would be prohibited or required? Would, for example, food grown by migrant workers for slave wages, or whose manufacture polluted rivers or involved extreme cruelty to animals be prohibited? The response they consistently gave was that while people in this community would be inclined to, for example, avoid such foods, there would be no community-wide laws governing their practices. People would do the right thing, presumably because they would be good people who always try to do the right thing within reason.[60]

Right. Just like the lawyers would presumably do the right thing without the antiquated ritual of an oath. I even find in my own writing process, for example, that I write more effectively when I've ritually set aside time for writing. Some writers do this

59 Ana Levy-Lyons, *No Other Gods: The Politics of the Ten Commandments* (New York, Center Street), 2018, 4-5.
60 Levy-Lyons, 3.

with a candle or ritual object, for others it may be just an item in their calendar, but creating a ritual container helps me to resist distractions, and hold myself to the sacred task. I don't always succeed in making these ritual boundaries. It's easy for me to feel the pull of the ideology of freedom and flexibility. "I'll just respond to these emails. I'll just read a few Facebook posts to relax…" Yet, when I do create a sacred time, a container—the writing usually flows more easily.

Rules and rituals that regulate our lives are the vessels and garments that contain our sparks and give them shape, allowing them to flourish. From seemingly small rituals like sharing family dinners, to communal rules about ethical purchasing as in Levy-Lyon's example, "externals" like rules, rituals and boundaries can be the paradoxical vessels and garments that give just enough form and protection to create real freedom.

This can apply to actual clothing as well. There are good reasons for wearing garments, even beyond keeping us warm in the winter. Clothing can cover, but it can also be a vehicle for expressing ourselves. My younger son, Eiden, for example, has since practically the time he was born, been interested in clothes. Don't ask me where he got it from—neither Ilana nor I are particularly fashion conscious. But when he was a toddler he would change his outfit up to ten times a day. Ilana needed to tell him that she would help him get dressed twice a day—once in the morning and once at night—and the other times he was on his own. He quickly learned how to dress himself. At five, he designed his own Jedi knight uniform which his grandmother sewed according to his instructions. Now, at 12, one of his favorite excursions on a day off from school is to go to the Cambridge Garment District and search for new items of clothing for his wardrobe: a rock and roll tee shirt, a leather jacket, an Indiana Jones hat…. he has a sense of style all his own. It is clear that he loves to express himself through clothing. And we all, to one extent or another, do so. Clothing is a means to express the inner self that we want to show to the world. They can also protect that inner self.

The *Pearl* and the *Flame*

The Ishbitzer Rebbe, Rabbi Mordecai Yosef Leiner (1801–1854), wrote in his work *Mei HaShiloch* about the meaning of the priestly garments. He writes about the breaches which cover the "private parts":

> ...every place in a person where God created the ability to build an eternal edifice (*binyan adey ad*)—there God created a possibility of loss, such that a person could lose everything. But through guarding oneself in this area, arises the ability to sprout from this place an eternal edifice.[61]

In other words, our genitals are especially appropriate to cover because it is from them that we create a *"binyan adey ad,"* an eternal structure, which here refers, on a surface level, to children or perhaps family in general. But in the specific usage of the Ishbitzer, *binyan adey ad* means more than physical progeny—it means a lasting contribution to something eternal. The family and children that one creates, in part through one's genitals, is meant to reflect trust, values, learning, and love. The holy vessel of the home nurtures those growing images of God.

That vessel does not need to look like the traditional home. But, a *binyan adey ad*—an eternal edifice, implies a balance between privacy and openness, enough of a vessel to hold and nurture the deep, mysterious and infinite mystery of a soul, but not too much to oppress and cover its light. Whether or not one is planning on making a baby, our psycho-spiritual knowing understands that there are places in our bodies that, as the Ishbitzer says, hold the possibility of creation, and also can be vulnerable points where we can lose our sense of self: as he says, "lose everything."

He's making a point about modesty which can be difficult to talk about because too often the idea of modesty has been used by traditionalists in oppressive ways, especially against women. However, in this contemporary moment of "freedom fetish"

---

61 Rabbi Mordecai Yosef Leiner, *Mei HaShiloch* (Benei Brak, Mechon L'Hotza'at Sifrei Rabbonaynu HaKedoshim Ishbitza Radzin), 1995, 89 (Vol. 1).

which says any rule is oppressive, it may be important to listen to the Ishbitzer's point that the most vulnerable, private and powerful parts of our bodies need care and protection.

I'm not in agreement with those who label Judaism, for example, a religion which is not "sex-positive" because there are rules in traditional Judaism regarding modesty and sex. While it's true, again, that those rules can be too rigid sometimes and can be oppressive, the existence of rules doesn't in itself mean that there is oppression. They could mean that the religion recognizes that this is an important area of life and therefore wants to give the best chance for success. Anthropologist Mary Douglas wrote about the rule and ritual heavy biblical book of Leviticus, "... it may be remarked that religions which ritualize sex are usually more in favor of it than against. To suppose that the numerous sexual regulations in Leviticus exhibit a narrowly puritanical attitude to sex would be like expecting a culture with numerous food rules to condemn good food."[62] As the Ishbitzer Rebbe says, in areas of great potential, there is also great vulnerability, which calls for a proper container.

These coverings, garments, and containers are all a part of the fractal, nested nature of our world. We don't exist all on the same plane, but we simultaneously are held in nested patterns of garments and sparks, vessels and light. Escaping these garments has an allure of freedom and liberation, but it can also harden us, make us defensive and paradoxically less open to diversity. Recognizing the nested nature of our world can help us in one of today's most vexing social and political problems: how to see past the divisions and ideological factions that are tearing our society and political culture apart. Like in composting where we learn that nothing is waste, we'll see in the next chapter how nestedness helps us learn to disagree, even argue with all our strength, and still get along.

---

62 Mary Douglas, *Leviticus as Literature:* (Oxford, Oxford University Press), 2000, 178.

For Heaven's Sake!
How to Argue and Still Love One Another

It follows from what I've said about nestedness and the need for some boundaries and coverings, that in our social and political life the goal shouldn't be for everyone to agree. The goal needs to be that we learn how to live with difference, even deeply consequential arguments, and still get along. We've seen in recent years what it looks like when this delicate balance falls apart: demonization and a fracturing of the democratic system.

Machloket L'shem Shamayim—is Hebrew for an "argument for the sake of Heaven." This phrase comes from the Mishnah, in Pirkei Avot (5:17), where it refers to the arguments between the Schools of Hillel and Shamai. It is said that they argued vociferously but that they still respected one another, would marry into one another's families and would teach each other's opinions. This is a reflection of the practice in the Talmud of teaching several opinions on an issue, not just the one that is considered correct. It's also at the root of the famous Jewish predilection for arguments, which spawned jokes like "two Jews, three opinions."

To connect this deeply Jewish idea to the natural principle of nestedness I want to come back to the garden, not specifically for composting, but the whole attitude of the gardener. A few years ago, I was struck by a paragraph that I read in one of Michael Pollan's earlier books, *Second Nature: A Gardener's Education*, in which he writes about gardening in a way that reminded me of Talmud:

> The gardener feels he has a legitimate quarrel with nature—with her weeds and storms and plagues, her rot and death. What's more, that quarrel has produced much of value, not only in his own time here (this

garden, these fruits), but over the whole course of Western history. Civilization itself, as Freud and Frazer and many others have observed, is the product of that quarrel. But at the same time, the gardener appreciates that it would probably not be in his interest, or in nature's, to push his side of this argument too hard. Many points of contention that humankind thought it had won—DDT's victory over insects, say, or medicines' conquest of infection disease –turned out to be Pyrrhic or illusory triumphs. Better to keep the quarrel going, the good gardener reasons, than to reach for outright victory, which is dangerous in the attempt and probably impossible anyway.[63]

Wow! I thought, Michael Pollan is talking about a *machloket L'Shem Shamayim* with nature. He recommends an environmental ethic that he calls the "Gardener's Ethic" in which we humans push our interests: we plant, build, weed and put up fences, but we don't try to vanquish nature. We understand, like the gardener, that we are in a perpetual "argument." And that is ok. It's actually the continued dialogue that is important.

The natural principle that underlies both the Talmud's and Michael Pollan's "argument for the sake of heaven," is the idea of nestedness. The gardener or the Talmudic rabbi, as the case may be, represents only one level. They are in disagreement on one level, but in unity on a larger, or deeper level. Both sides in the argument are not arguing purely for their own interests, but they place their interests into a larger context: Heaven or Nature.

In Pollan's case the gardener lives on at least two levels. She knows that vanquishing nature would be like drilling a hole in her own life-boat. She has her own interests as a human who wants to see her tomatoes grow. That is one level—the human community and its interests But the good gardener realizes that our human community is nested within a larger community of the natural

---

63 Michael Pollan, *Second Nature: A Gardner's Education* (New York, Grove Press), 1991, 193.

world. I may have a local argument with the weeds in my garden, but destroying nature (the complex health of the soil, the ecology of insects, birds, etc., which keep pests in control, and so on) in the process of controlling these weeds would ultimately be self-defeating. The good gardener realizes that the argument is only on one level, but both sides are unified on a higher level. We are all nested in our ecosystem, and beyond that, in the world of nature.

The Jewish tolerance for argument is related to the idea of Jewish peoplehood: that very nestedness by which Judaism sees a set of concentric circles starting from the individual, the family, the Jewish people/nation, the larger human world, and the world as a whole. Our arguments take place within a nested context of a larger whole.

While Judaism is a religion, it does not define itself as a religion in the classic sense with which we are familiar in the West where we often use "faith" as a synonym for one's religion. In Christianity, at least in theory, a person is only a Christian if they believe in Jesus as the Son of God. The idea of an atheist Christian is, at least in theory, an oxymoron. But atheist Jews don't upset any Jewish cognitive maps.

That is because Judaism has defined itself as a people—true, a people who have carried a religion, a set of ideas, or beliefs—but even if you don't happen to believe in that religion, set of ideas or beliefs, you're still a Jew. In terms of nestedness you could say that a Jew's Jewish ideas and beliefs are on a more superficial level than that Jew's Jewishness. Because Judaism is inherited by birth, or by a conversion ceremony that uses the symbolism of a new birth (the mikveh, or ritual immersion, which is the *sine qua non* of Jewish conversion, is often compared to a symbolic return to the womb) Jews are Jews in their bodies. In many ways, this embodied Jewish identity is a deeper level than thoughts. Being a member of a family or tribe creates an emotional attachment and sense of identity which frequently is felt as deeper than thoughts or opinions, which may change many times over a lifetime.

The *Pearl* and the *Flame*

So, when Jews disagree, it really is like a family argument. I know from watching my kids that they feel a lot more freedom to argue and even yell and scream and act like brats with their mother and me. But with other people, they are little angels. That is because they know that we are not going to leave them or kick them out because of an argument or bad behavior. We are family. There is a deeper connection that encompasses and holds the disagreement and makes it safe to have. That is why we have two Jews and three opinions.

This also relates to the ancient Jewish practice of arguing with God. It goes back to Abraham, who protested when God let him know about God's plans to destroy the cities of Sodom and Gomorrah. Abraham tossed God's own ethical standards back in God's face: "Should the Judge of the World not do justice? Would you destroy the innocent with the guilty?!" [Genesis 18:25] God, by the way, was probably telling Abraham about those plans in order to get him to argue. In many cases, the biblical God was quite anthropomorphic, and is found to be angry, sad, regretful, loving, forgiving—all the human emotions. So, when God was angry, God relied on human partners, prophets like Abraham, to keep that anger under control.

The tradition continued with famous examples such as Rebbe Levi Yitzhak of Berdichev, an early Hasidic master who was famous for convening a *beit din*—a court of religious law—on Yom Kippur in which he put God on trial for not holding up God's side of the Covenant with the Jewish people. We, he argued, have been doing our job, keeping the mitzvot under the most difficult of circumstances of poverty and oppression, and all You can do is send disease, wars and misfortune?!

In the most moving of examples, we have the testament from within the Warsaw Ghetto of Rebbe Kalonymous Kalman Shapiro, the Hasidic Rebbe of his community in the Ghetto, who we have cited earlier. He continued to lead his people, and wrote down his sermons throughout the war years. When he saw that he, too, would soon be deported, he buried his writings in a metal container. They were found by a Polish construction worker in the 1960s and sent, according to the instructions on the container, to his family in Israel. He writes, for example:

> We trust that You will save us and that You have not forsaken us completely, heaven forfend, but in this respect You have forsaken us—with respect to the fact that "[thou] art far from my help"—that the salvation is so long in coming and the sufferings have dragged on for such a long time.
>
> … How can You tolerate the humiliation of the Torah, and Israel's anguish? They are being tormented and tortured just because they fulfill the Torah![64]

Nehemia Polen, who selected, translated and provided commentary on his ghetto writings in his book *The Holy Fire: The Teachings of Rabbi Kalonymus Kalman Shapira, the Rebbe of the Warsaw Ghetto*, notes that Rabbi Shapiro himself commented on the proper way to argue and protest to God:

> Now if the Jewish person speaks this way as an expression of prayer and supplication, as he pours out his heart before God, that is good. But if, God forbid, he is posing questions; or even if he is not [actively] questioning, but, in the depths of his heart, his faith, God forbid, weakened, then God help us![65]

Polen explains, "In other words, expressions of protest and challenge are quite proper when directed as part of an ongoing relationship with Him." That is to say, when one stays within the boundaries of the relationship, then conflict and disagreement are possible—because you haven't put the relationship itself into

---

64 As translated in Nehemia Polen, *The Holy Fire: The Teachings of Rabbi Kalonymus Kalman Shapira, The Rebbe of the Warsaw Ghetto*, (Lanham MD, Jason Aronson), 1999, 100.
65 *The Holy Fire*, 101.

question. Your argument is nested within a larger framework. It is only when you step outside of the container of the relationship to question that container itself that the questioning becomes destructive and dangerous.

Though there is a giant chasm between the relationship between God and the Jewish people during the unspeakable horror of the Holocaust and the mundane arguments that occur within a marriage, both follow the same dynamic of relationship. So, it is not surprising that the advice of leading relationship researcher John Gottman in his bestselling book, *The Seven Principles for Making Marriage Work,* is essentially the same advice as the Piaseczner gave about arguing. He says that when couples speak to each other in ways that complain about a specific action, that is fine. When they go beyond specific complaints to blaming the person and questioning the relationship, that is a bad sign for the marriage.

Ilana, for instance, tends to leave her tea, with the tea bag and milk still in it, so that after a couple of hours on a hot day the milk curdles and smells, wherever she happens to put it down. Does this drive me crazy? As you can imagine, yes, it does, and I complain about it to her regularly. I, on the other hand, have the perfectly harmless habit of tending to leave whatever project I'm working on, such as trying out different frames for some unframed paintings, laying around the house for weeks until I finally come back to either finishing the project or giving up on it. Does that drive her crazy and does she complain to me about this habit? Yes, she does! But we're still happily married and, while these complaints are irritating in the moment, they don't come close to the core of our love for each other.

When couples, instead of specific complaints such as ours, turn to blaming their partner, using sarcasm and personal insults, that undercuts the relationship itself, endangering the container which makes arguments safe and even productive. The same dynamic applies to our politics and society, as we'll see in the next section.

Disagreeing Today

In our society today we have a crisis of civil discourse. This is certainly a result of social media which greatly amplifies disinformation and conspiracy theories. This same online business model also tends to put people in social and political silos, as internet algorithms feed people the information that they want to hear, and keeps them in a bubble of like-minded people.[66] Some have also traced it back to the '60s counterculture questioning traditional measures of truth and to academic theories which have put objectivity into question. In my eyes much of the counterculture and academic theorizing have been positive contributions to our culture but once the genie of a more complex and subtle understanding of truth is out of the bottle we've seen the breakdown of standards of truth spread to every corner of the cultures including right wing talk radio, cable T.V. and more. Whatever the cause or causes, we find ourselves in a polarized society where civil discourse almost doesn't exist anymore. We find ourselves caught within what Gregory Bateson many years ago called "schismogenesis."

He was doing anthropological research in New Guinea and produced the classic anthropological book, *Naven*, where he coined the term schismogenesis to describe the feedback loop of increasing fragmentation when two conflicting groups within a tribe start to argue. The argument grows and grows until the whole society falls apart. It was one of the first social scientific applications of what we'd call today complex systems theory. We need arguments but not the kind that we have now. Parker Palmer writes in *Healing the Heart of Democracy*,

> I neither imagine nor yearn for a conflict-free public realm, a fantasy that is tantamount to yearning for a death-free life. Only in a totalitarian society is conflict "banished." Conflict does not disappear, of course, but is merely driven underground, replaced with

---

66 See, for an excellent exploration of this issue, the 2020 documentary, *The Social Dilemma*.

a public illusion of unity that must be enforced by violence. In a healthy democracy, public conflict is not only inevitable but prized. Taking advantage of our right to disagree fuels our creativity and allows us to adjudicate critical questions of many sorts: true versus false, right versus wrong, just versus unjust.[67]

I'm reminded of what Talmudic scholar David Kraemer says about why the Babylonian Talmud so thoroughly embraces argumentation.[68] He notes that the rabbis of the Talmud paradoxically realized that though they were the guardians of a divine revelation, they themselves were not prophets. They were humans and all they had were the texts left behind by prophets such as Moses. Therefore, all they could do was interpret those texts. That meant embracing argumentation, and the Babylonian Talmud became that rare holy text which almost never gives an answer, but instead takes the reader through the thought processes of argumentation. Even though we don't have the Truth, the ancient rabbis reasoned, we can sharpen our understanding through arguments, and come a bit closer to the truth.

Kraemer quotes the philosophers Chaim Perelman and L. Olbrechts-Tyteca (*The New Rhetoric: A Treatise on Argumentation*) in saying that argument is necessary for civil society: "Only the existence of an argumentation that is neither compelling nor arbitrary can give rise to human freedom, a state in which a reasonable choice can be exercised."[69] That is, if the choice is between a mathematical proof that no reasonable person could argue against, or merely arbitrary ways of making decisions—such as violence or propaganda—most questions that a society faces will fall into the arbitrary category. We need to be able to

---

67 Parker Palmer, *Healing the Heart of Democracy: The Courage to Create a Politics Worthy of the Human Spirit* (San Francisco, Jossey-Bass), 2011, 61.
68 David Kraemer, *The Mind of the Talmud: An Intellectual History of the Bavli* (Oxford, Oxford University Press), 1990, "The Meaning of Argumentation," 99-128.
69 *The Mind of the Talmud*, 113 (quoting *The New Rhetoric*, 514.

talk to one another, understanding that we have some things that hold us together, and other things that separate us. We could use some of that today.

Sh'ma—Listen to our Heart

During the month of Elul, the last month of the Hebrew calendar, usually in August or September, right before we get to the High Holidays, many Jews have the custom of saying Psalm 27. It contains this line, "For Your sake, my heart said, 'seek My Face,'—I will seek Your face." God speaks through our hearts. As in the Sh'ma with which we started this section, we need to start from the heart and work our way up through all the nested parts that make up the whole.

In *Healing the Heart of Democracy*, Parker Palmer similarly focuses on the importance of our inner lives, our hearts, in our politics. He writes:

> In this book, the word heart reclaims its original meaning. "Heart" comes from the Latin *cor* and points not merely to our emotions but to the core of the self, that center place where all of our ways of knowing converge—intellectual, emotional, sensory, intuitive, imaginative, experiential, relational, and bodily, among others. The heart is where we integrate what we know in our minds with what we know in our bones, the place where knowledge can become more fully human. *Cor* is also the Latin root from which we get the word courage. When all that we understand of self and world comes together in the center place called the heart, we are more likely to find the courage to act humanely on what we know.[70]

What he describes is an ancient understanding of the heart, which we've mostly lost in modern times. We speak of the center

---

70 *Healing the Heart of Democracy*, 6.

of knowing as the brain, but the biblical use of heart agrees with Palmer's definition: when the biblical verse of the Sh'ma says "place these words upon your heart" it is speaking about that central, integrated knowing place which is much deeper than the brain. Palmer continues,

> "The politics of our time is a 'politics of the brokenhearted'—an expression that will not be found in the analytical vocabulary of political science or in the strategic rhetoric of political organizing. Instead, it is an expression from the language of human wholeness."[71]

His connecting broken heartedness with human wholeness reminds me of a famous Jewish saying attributed to the 19th century Polish Hasidic master, Rabbi Menachem Mendel of Kotzk, known as the Kotzker Rebbe (1787-1859): "there is nothing more whole than a broken heart."

When Palmer talks about broken-heartedness, he points to two possibilities: a heart that is broken open or a heart that is broken apart. Much of what we see in our country of rage and demonizing immigrants or opposing political parties are manifestations of hearts broken apart. Palmer says, "...rage is just one of the masks that heartbreak wears."[72] Yet, when we can examine, talk about, and let ourselves feel our broken hearts, they may break open, letting us feel more compassion and love, to hold the tension of disagreement and difference as a creative tension.

Our ability to argue and accept differences is a matter of expanding our vision to our larger vessels: beyond our political party to the nation, beyond our nation to the world, without letting go of our more local vessels. It is also a matter of taking the most interior nested layers seriously, the sanctuary of our hearts is as necessary as the work of our hands or our influence on

---

71 Ibid.
72 Ibid.

large corporations or nations, in creating a diverse and flourishing world.

The conflict that I felt when I was pulled into Aish HaTorah so many years ago was the result of a world split apart: though there were many attractive aspects to their deeply pious form of Judaism, they were operating in a paradigm of us against them, Torah versus modernity. I couldn't accept that split, in the world or in myself. Luckily, the deeper I dug into Judaism, the more I noticed that this "us against them" mentality wasn't reflected in the core texts, at least as I read them. There was a much more nuanced way of distinguishing my group, or my opinion, from others': argument for the sake of heaven—a way to disagree but also acknowledge our connection. I also saw that our coverings, vessels, and divisions aren't necessarily standing in the way of Oneness, but can be the needed containers that lead us toward unity.

And, I also saw that the modern world was coming around to a similar way of thinking. Reading Michael Pollan and his gardener's ethic was one of the openings to seeing a shift in our contemporary paradigm toward acknowledging our nestedness, seeing the many sanctuaries starting from our hearts, and including all various boundaries: bodies, families, community, nation, human-kind, and earth.

Whether I'm talking about creating a civil culture of productive disagreement (something that we could definitely use today, especially in politics), or whether I'm trying to find a good way to both protect nature and also make the world comfortable and hospitable to us humans, thinking in terms of nested sanctuaries provides a framework that works.

"Intractable paradoxes" that still plague us every day like the individual versus the community, or universalism versus particularism, disappear when we shift our mindset. When we see ourselves in nested levels of reality we honor all parts of our being as sanctuaries and exclude none. Taking things seriously on our own small, individual level, feeling that it actually does make a difference if I do something or not, really can change the world. In the next chapter, we'll explore precisely why those small actions can make all the difference in your life, and in the world.

PART FOUR:

Mitzvah
The Spiritual Technology of Change

For want of a nail, the shoe was lost;
For want of a shoe the horse was lost;
For want of a horse the rider was lost;
For want of a rider the battle was lost;
For want of a battle the kingdom was lost!

— Folk proverb quoted in James Gleick's *Chaos: Making a New Science*

"Be very diligent in a small mitzvah *as in a large one; you never know the reward of a* mitzvah."

— Pirkei Avot 2:2

"Hope… is not the same as joy that things are going well, or willingness to invest in enterprises that are obviously headed for early success, but, rather, an ability to work for something because it is good, not just because it stands a chance to succeed."

— Vaçlav Havel

Mitzvah and the Tipping Point

Sitting in her apartment in Cambridge, MA., Ilana pondered how she could get to her friend's wedding. The wedding would take place the next Saturday night, a four-hour drive away, in New Paltz, New York. That wouldn't have been a problem except that, as someone who observed Shabbat, including not driving in a car on Shabbat, to get to the wedding on Saturday night she would need to find a place to stay in New Paltz for the entire Shabbat, from sundown Friday night all through Saturday. The logistics were clearly a hassle, and, though she felt a heart connection to the people getting married, they were fairly new friends at the time. She didn't quite understand why she felt strongly that she needed to go to this wedding.

Shoshana, a friend visiting from Israel, happened to be staying with Ilana. They were brainstorming these wedding logistics when Ilana mentioned, "You know, there is this guy, Natan Margalit, who lives in New Paltz, but I hardly know him at all. It would be weird to call him out of the blue and ask to stay for the whole Shabbat."

Shoshana responded, "I know Natan! He lived in Israel for years. In Israel, dropping in at the last minute is completely normal. Unexpected guests are the spice of Shabbat. He'll be totally fine with it. Call him."

Ilana called me up out of the blue and, of course, I told her it was fine for her to stay for Shabbat. Then, I got on the phone and started calling around for other guests to come for one or two of the Shabbat meals. I didn't have anyone else coming over yet, and I didn't want Ilana, whom I barely knew, to feel uncomfortable, spending the entire 24 hours of Shabbat alone with me. She wasn't interested in a date, after all. She just needed a place for

Shabbat. But, call as I might, I only found one other person who could join us for lunch.

So, Ilana and I ended up spending just about the whole Shabbat alone together. We took a long walk on the rail-trail, meditated, ate my homemade Challah (which she loved) and my tuna pasta dish (not so much). We went to the wedding together since the groom was also a friend of mine. It was when we were dancing together that something almost literally went "click." I couldn't describe what, but something special was happening. Then, like Cinderella, she had to get a ride back that night to Cambridge and was gone. Luckily, I had her number.

I called her on Sunday morning to see if we could arrange to see each other again. She told me that on her ride back to Cambridge, she had said to her friend, "I just met the man I'm going to marry." I was a little taken aback, but didn't disagree. I just suggested that we should date a bit before deciding anything. Five months later we were engaged, and five months after that we were married.

Small Actions, Big Results… Maybe

One small action, like deciding to go to a wedding, can have truly miraculous, life-changing consequences. Or, it might not do a thing. You never know. We are getting to the heart of the shift in thinking that I'm calling Organic Torah. When a small action can have wildly out of proportion results, it means that there is openness built into our world. That fact has been a slap in the face to the more overconfident proponents of modern, reductionist science. This element of openness threw the scientific world for a loop in the early 1960s, when Chaos Theory first emerged. If we can't predict it, it must be chaos!

Yet, as the scientists who first formulated Chaos Theory found out, while individual events will always escape our abilities to predict and control, regularities and patterns can be discerned on the level of the whole system. Oxymoronic as it sounds, "Chaos Theory" understands that there is order in chaos. Not only that, but this gap between an action and its consequences has implications far beyond science: it is a spiritual key, discovered simultaneously by different religious cultures over two thousand years ago, that opens up a world of freedom and holiness.

Letting go of the fruits of one's actions and turning them over to God is the source of the religious system of dharma in Hinduism, as seen in the spiritual classic, the *Bhagavad Gita*. In the *Bhagavad Gita*, the god Krishna, disguised as the prince Arjuna's chariot driver, instructs Arjuna to go into battle simply because it is his duty, his dharma.[73] He must give up on focusing on the fruits of his actions, dedicate those actions to Krishna, and then simply do the action. This book has been interpreted, most famously by Mahatma Gandhi, in the metaphorical sense: the battle is our lives, not an actual military campaign. We must face each challenge in our lives by doing our duty (dharma), simply because it is the right thing to do (or, the will of God), and let go of the fruits of our actions. This transforms daily life into an arena of holiness.

This combination of small actions and uncertainty is also the basis of the rabbinic system of the mitzvah. From about 100 to 500 CE, during the Roman Empire (and about the same time as the *Bhagavad Gita* was being written in India) the Rabbis of the ancient Mediterranean were developing what might be called the basic spiritual technology of Judaism. It's a system of small actions which are usually understood as commandments from God.

Spiritually inclined interpreters will often use a Hebrew/ Aramaic play on words to connect *mitzvah* to *tzavta* which is Aramaic for "together." This little word play/interpretation emphasizes that it is through small actions that one comes together with the Divine Source. Whatever way we understand it, the reason for doing the action is above and beyond one's own self-interest. How it all comes out is not necessarily the main

---

73 This is one of the various shades of meanings for the Sanskrit word dharma. It can also mean created order, reality, law and more. It is also used differently in Buddhist, Hindu, Jain, Sikh and other contexts.

point. Rather, it is experienced as a worthy action in and of itself. It connects a person to a higher purpose, or, as the ancient rabbis put it, doing a mitzvah purifies the heart.

Mitzvah: The Rabbi's Spiritual Technology

In common speech, people talk about "a mitzvah" as a good deed, a nice thing to do. "Do a mitzvah," the Jewish mom says to her child, "give that man on the street a dollar." Yet, in traditional Judaism, a mitzvah is the building block of a complex legal system. A mitzvah might be a "good deed" like giving a homeless person some money. It also could be refraining from working on Shabbat, or, it could be listening to the sound of the ram's horn Shofar on Rosh HaShana. The rabbinic framework of mitzvot (the plural of mitzvah) covers all aspects of life: moral, ethical, ritual, business, and pleasure. It is a legal system attempting to describe, and induce people to follow, a complete way of life. *Halakhah*, the Hebrew word for this legal system, literally means "the path."

One second century rabbinic sage said, "Be very diligent in a small mitzvah as in a large one—you never know the reward of a mitzvah." (Pirkei Avot 2:2) In other words, do the right thing now—you never know how even a small action might affect the big picture. So, as they say, just do it. We may find this idea hard to swallow. It may be counterintuitive to think that a small deed can change everything, but folk wisdom has been clear on this point. It is the stuff of folk tales across the world, and the Jewish tradition has its fair share.

From the time of the Talmud comes a legend about the daughter of the greatest sage of his generation: Rabbi Akiba. She was, according to the tale, fated to die on her wedding day (based on Bablyonian Talmud, Shabbat 156b). All the astrologers said so, and Rabbi Akiba was worried. And, in fact, on her wedding day, a poisonous snake hid behind the wall of her room, waiting to strike her. But when she came back to her room and undressed, she happened to take off her hair pin and stuck it into the wall. It went through the wall, into the snake's head, killing it, and saving her life.

Later, when they found the dead snake, her father asked her if she had done any mitzvah during her wedding festivities. She answered, yes: during the wedding feast there was a poor beggar who came to the door of the wedding hall. No one was paying him any attention, but she got up from her place of honor and gave him her plate of food. Rabbi Akiba said, "My daughter, that mitzvah saved your life."

This story is couched in the folktale language of supernatural intervention, but it illustrates a characteristic of complex systems: a small act can make all the difference. It could save your life. In our contemporary culture, this is the key to the popular writer Malcolm Gladwell's ability to come up with deliciously surprising and counterintuitive stories in his best-selling book, *The Tipping Point: How Little Things Can Make a Big Difference*. Gladwell never mentions complex systems, but the whole book is based on this counterintuitive—yet folktale familiar—quality of complex systems: small actions can have big effects—sometimes. He popularized the phrases "tipping point" and "going viral," which are signposts to an emerging cultural shift toward thinking that is less mechanical, more open to the surprising jumps and twists of living systems.

Small actions can affect the world in positive or negative ways. From the Hasidic tradition, which started in the 18th century in the struggling Eastern European Jewish communities, comes a cautionary tale which looks at the consequences of missing a chance to do a mitzvah.

There lived a man named Reb Dovid, a Hasid, which means a follower of a revered Rebbe, or Spiritual Master. As the story goes, this Hasid was on his way to see his Rebbe for the holiest day of the year: Yom Kippur. This pilgrimage was the highlight of his year and he had made every preparation for his trip. Not far from his Rebbe's town, he passed by another, much smaller, village. As he approached the village, he saw a group of Jewish men gathered on the roadside. They hailed him and pleaded, "Please, can you stop and be with us for the festival to make a minyan?" (A minyan is ten Jewish men to make the traditional quorum for prayer.)Reb Dovid was in a quandary. To join these people who needed him for a minyan would be an important mitzvah. They couldn't fully celebrate Yom Kippur without a minyan. Yet, he'd waited a year to be with the Rebbe! There was no comparison. These were simple villagers. He'd never hear the deep wisdom flowing from the lips of his Rebbe in this small village; he'd never be enchanted by the singing and the intensity of prayer at the court of his Rebbe, here with these ignorant householders."I'm very sorry, but my answer must be no," he said, as he spurred his horse and rushed off.

When he got to his Rebbe's village and tried to greet his Master, the Master wouldn't look at him. That's okay, he said to himself, the Rebbe has a lot of people vying for his attention. I'll greet him in the morning. It was the same thing in the morning, and the afternoon. The Rebbe looked right through him as if he didn't exist. Finally, just as the holy day was closing, he couldn't restrain himself. He threw himself at the feet of his master and cried, "What have I done? Why haven't you greeted me this whole time?"

His master looked at him with eyes filled with sadness and compassion. "Reb Dovid, your soul had been waiting a thousand years to join those nine men in that village. All our prayers and songs were nothing compared to the spiritual fulfillment waiting for you with them. Now, your soul may have to wait another thousand years."[74]

We've all been there on that "detour," that unexpected invitation which looks like it's taking you off your path, but in fact it is your path. Kurt Vonnegut once said, "Peculiar travel suggestions are dancing lessons with God."[75] Back in Jerusalem, when Rabbi Meir Schuster invited me to the Old City to attend a "class on Jewish philosophy," I was on my way to an Israeli folk dance festival. I had no idea what this little detour would mean for

---

74 See Shlomo Carlebach, *Shlomo's Stories: Selected Tales,* (Lanham, MD, Jason Aronson), 1996, "A Meeting on the Road," 53-58,
75 From *Cat's Cradle.*

my life. Or, there was the time I got an unexpected invitation from my friend Azriel to help him lead a Passover Seder in Dharamsala, India. I knew it would be an interesting experience, but I never guessed it would lead to me meeting a woman there who would be my girlfriend for three years, introduce me to meditation, Eastern religions, Sufism, and a whole new perspective that would change my life.

Some theoretical physicists say that each decision creates an entire alternative universe. That may or may not turn out to be true. Yet, from the wisdom I've gathered, I can say that at any moment, God, or the universe, or your destiny is asking you a question. It is up to you to answer. You can connect that moment to the Eternal, or you can let it fall away. Never mind all the books you've read, never mind your own ideas for the future or your plans—the only question is: what is God asking of you in this moment? That is the true mitzvah.

It isn't always easy to discern the answer, and that is one of the reasons why religions have guidelines, laws and custom, long periods of study, and spiritual practices—sometimes so much that they feel like they get in the way of acting spontaneously in the moment—but, ultimately, all that training is for the purpose of being able to discern: "what is the right action, right now?"

Tools for Discernment

This quality of reacting to the present moment, of course, is central to meditation and the Eastern spiritual traditions that have become popular in the West. Last year, my wife Ilana went to a silent meditation retreat. During one of the instructor's talks, there were a few minutes for questions. A newcomer to silent meditation asked the teacher, "If I were to do this practice for twenty years, what would I gain?"

The teacher answered, "You would be more likely to respond appropriately to whatever situation you're in."

Maybe it's that simple. The goal is to respond appropriately to the question of the moment. That means letting go of my story, my childhood narrative, the baggage that I carry with me into the situation and that constantly gets in the way of simply responding appropriately.

Responding appropriately to the moment can be life-changing, or simply make for a better day at home. Not long ago, Ilana bought some dark chocolate, 88% cacao. I like a little less cacao because I'm sensitive to caffeine. What was she thinking?! Doesn't she care about me?! How can she be so insensitive?! All these responses went through my mind before I caught myself (luckily, this time) and realized that this had a lot more to do with my "stuff": the deeply unconscious effects of my relationship with my mother and father, my childhood rivalry with my siblings, my mean baseball coach in junior high, and who knows what else. All my baggage.

It probably isn't appropriate at this moment to get angry at my wife for buying the chocolate. When I let go of the baggage, I'm much more likely to respond appropriately. I'm much more likely to do the mitzvah I'm called upon to do, which is to thank her for buying the chocolate.

From Ancient to Modern
The Beginning of Chaos

This ancient spiritual technology, the mitzvah, embodies the wisdom of what they'd call in complex-systems lingo "bifurcation events." Or, in the language that's been made popular by Malcolm Gladwell, tipping points. The old proverb, "the straw that broke the camel's back," works just as well. It is this space of indeterminacy in a complex system that makes for so much of the confusion, but also the excitement and hope of our world. On any given day, the underdog might beat the favorite. Every wave I surf is unique, unpredictable. A small group of dedicated people can, and do, change the world.

You may see the statistics that people from your neighborhood don't go to college, but that doesn't guarantee you will not go. Your candidate may be unlikely to win, but you never know, she could. David can beat Goliath. My blog post can go viral. Taking out my compost or speaking up at a meeting can make a difference.

This uncertainty factor was the beginning of Chaos Theory. Back in 1961 Edward Lorenz, a researcher in climatology at MIT, made a discovery that would set science in a new direction. He was using early computers to attempt to create a model by which we could predict the weather. We can, after all, predict an eclipse or the return of a comet pretty accurately over hundreds of years. With the new technology of computers, many scientists were excited that perhaps we could now do the same with weather.

Lorenz was working on one of his computer models of a weather system. One day, for some reason, he needed to stop the computer and re-enter the coordinates on his experiment. He rounded off one number, 0.506127, to 0.506. According to the prevailing norms of science that tiny difference shouldn't have made any difference. Science then operated on the well-established principle of "proximate knowledge of initial conditions." That is, a slight change in the initial conditions shouldn't make much of a difference in the end result. He was shocked when he returned a few hours later to find that the results were wildly off what they were before. The whole principle of "proximate knowledge of initial conditions' ' was called into question.

This discovery and others like it became the basis of the famous "butterfly effect." This (hypothetical) example said that if a butterfly in China flaps its wings, causing a tiny ripple in the air currents, that ripple could set off a series of effects, which could cause a hurricane in Kansas. Thus, "Chaos Theory," was born, so named because it seemed to scientists that they were dealing with chaos.

In fact, it wasn't chaos. We know that there are patterns in complex systems. We can't predict the individual events, but we can predict the larger-scale trends. When economists study a particular economy, they can predict in general what percentage of people will be middle-income, how many below the poverty line, how many in the upper income brackets; but we can't say with certainty which individuals will occupy those places. It is the same with climate change: we know that the warmer temperatures will cause more severe storms, but we can't say exactly where or when a particular storm will occur. In traditional Jewish law, *halakhah*, we can know the principles, but we can't know how to judge the mitzvah of any particular moment until that case arises. We can show averages and tendencies on the scale of the whole, but there is always indeterminacy on the scale of the individual.

It was this butterfly effect, the imposition of uncertainty into science that pushed scientists to recognize that a paradigm shift was necessary. At first it seemed to be chaotic. But soon it became clear that chaos theory is not chaotic; it's just complex.

"Paradigm Shift" as Paradigm Shift

This aspect of uncertainty within complex systems brought science, during the '60s, '70s and continuing on until now, to explore and account for paradigm shifts. If science must account for uncertainty, if change isn't linear but happens in sudden jumps—that goes against some of the basic assumptions of scientific thought. So, it is perhaps not coincidental that when the historian and philosopher of science Thomas Kuhn proposed that science itself moves in nonlinear, sudden shifts, his ideas were not met with open arms.

In his 1962 book *The Structure of Scientific Revolutions* Kuhn proposed that "normal science" goes along basically filling in the blanks of an accepted set of assumptions, a paradigm. Most science builds up the edifice of knowledge brick by brick, one piece adding to the next. But every so often scientists come up with ideas that are outside of the accepted paradigm. These scientists have their papers rejected by peer-reviewed journals and they are passed over for the best academic positions. Their work is seen as irrelevant or heretical. Yet, sometimes, these outliers turn out to be right, and their work eventually shifts the scientific world, and sometimes the rest of the world as well, into a new paradigm.

The most famous of these paradigm shifts is probably Copernicus shifting our worldview to see that the earth revolves around the sun and not the other way around. For centuries astronomers had been developing an ever more cumbersome system of "epicycles" to try to explain the movement of the planets and stars under the prevailing earth-centered theory. Copernicus, Galileo, and others were persecuted and ridiculed for their heretical ideas, but they ended up bringing about a paradigm shift that went well beyond science and changed the way we think about ourselves and the universe.

Just as Edward Lorenz's Chaos Theory didn't sit well with most scientists at first, Kuhn's picture of science wasn't an immediate hit. It didn't look *scientific* enough. It seemed to depend on subjective factors like point of view and assumptions; it made knowledge jump from one state to another rather than moving linearly forward. In other words, Kuhn's idea of Paradigm Shift was itself a part of the current paradigm shift away from linear, reductionist science and toward complex-system science. Things

177

jump. Change can be sudden and unpredictable. People were uncomfortable with this. They asked, "Where does this leave us? Is this science or religion?"

The Talmud, Uncertainty, and the God of the Gaps

It is human nature to seek answers to the many puzzles and mysteries that life presents us. I often hear the claim that ancient, pre-scientific, peoples simply inserted God (or the gods) as a ready answer to those puzzles and mysteries.

"A good year on the farm?"

"God did it!"

"Feeling sick?"

"The gods did it!"

But God's role wasn't always simply obvious, even in the ancient world. Like many of us, our ancestors also pondered the question: Where do we see God acting in the world?

Even in the Talmud people asked this question. As is typical in the Talmud, the rabbis get at large theological issues through the unlikely avenue of small, seemingly technical questions of daily life and law. In this case, which appears in the Talmud (Sanhedrin 102b) a mundane, technical question of ritual was asked: "To what place on the bread is one supposed to point when one says the blessing over the bread before eating?" The answer given is that one must bless the bread right at the place where it first starts to form a crust.

Okay. But this wouldn't be Talmud study if we didn't ask: why? What's the logic behind this answer that we should bless the place where the crust starts to form? A 16th-century commentator, Rabbi Judah Loew ben Betzalel (1520-1609) known as "the Maharal of Prague," suggested this explanation: At first glance, we seem to do all the work of making bread. We plow and plant, harvest and thresh, grind and bake. Where was God in this process that we should bless God, saying "Who brings forth bread from the earth?" Didn't we do all the work ourselves? The answer is that when you put the bread into the oven there is no way that you can tell exactly where on the bread a crust will begin to form. There is something, in other words, about the process that they couldn't explain, and this mystery points to a power beyond us.[76]

But, wait a minute! This looks like one of the weakest arguments for the existence of God: it's called "the God of the Gaps." People will find something that we can't explain and then say, "There, you see, there must be a God!" It's not a strong argument, because if we are basing our belief in God on these areas we can't explain, in other words, if God lives in the gaps in our knowledge, then the place for God keeps getting smaller and smaller until there is no place at all. This comes up all the time in the history of science and religion. Science keeps coming along and explaining things: how all the species got here, how the world was formed, and fewer and fewer people have a need for God.

Is that really the argument that the Talmud is making? I don't think so. I think we can find out about the Talmud's answer by looking at some of the more recent trends in science.

Since the middle of the 20th century, there has been a trend in science that has not filled in more gaps but rather has shown us that there are things we can never know. In 1927 Werner Heisenberg published his Uncertainty Principle showing that it is theoretically impossible to know everything (or at least to know everything at the same time) about the movements of subatomic particles. In 1931 Kurt Gödel published his mathematical Incompleteness Theorem which proved that there will always be true statements that cannot be proven within any logical system. We might view Thomas Kuhn's theory of "paradigm shifts" as a part of this trend: we can't predict from the trajectory of "normal science" where the next great breakthrough will be. More recently we have seen the emergence of complex-systems theory. This says that in a complex system, a system with internal feedback—which includes a lot of what we experience in the world: weather, social trends, ecosystems, economic systems, and many more—there is

---

76 See *Sefer Netzach Israel* (Israel,Henig and Sons), 1980, 15-19.

no way we can predict what an individual in that system will do.

So, at least some parts of science have been coming back to an awareness that I think the Sages of the Talmud were talking about in their discussion of the crust of bread: there are things in this world that we can never control or predict. There is an essential openness, freedom, and mystery built into the structure of the world. This is not a failure of our science: there simply are gaps that will never be filled. When the Talmud says to bless that spot where bread first forms a crust, it tells us to honor that mystery, have some humility and realize that we can't know everything. It asks us an essential question: In those places in life with no explanation, what do we do with it? How do we interpret it?

This reminds me of a joke: There's a guy driving his car in New York City. He has a very important meeting to get to and he can't find a parking space. Desperate, he starts praying to God. "Please, God, find me a parking space! I promise I'll go to synagogue every Shabbat. I'll give ten percent of my income to *tzedakah* (charity)!" Suddenly, a car pulls out right in front of him and there's his parking space. He quickly says, "Oh, never mind, God. I found one."

This usually gets a laugh. It also makes an important point. It doesn't matter what the outside "evidence" or circumstances are: it is up to us to interpret the events of our lives. We can find miracles or we can find coincidences. We can find a Greater Consciousness or we can find our ego. There are always opportunities to exercise our free choice and serve a higher calling. In our lives, we can see the parking space open up and say, "Never mind, I found one." Or we can marvel at the mystery and perhaps whisper a quiet thank you.

These recent trends in science, trends that acknowledge absolute limits to our knowledge, open up a space for a different kind of knowledge. Science can help us know some things, but there is a realm of mystery upon which science can't shed light. This was a central theme in the writings of one of the most important Jewish voices in the twentieth-century, Rabbi Abraham Joshua Heschel. Heschel writes:

The search for reason ends at the shore of the known; on the immense expanse beyond it only the sense of the ineffable can glide. . . We do not leave the shore of the known in search of adventure or suspense or because of the failure of reason to answer our questions. We sail because our mind is like a fantastic seashell, and when applying our ear to its lips we hear a perpetual murmur from the waves beyond the shore."[77]

We have gotten so used to being in the world of reason, of knowledge, to the point where we have dismissed the realm of the ineffable as simply darkness and ignorance. We have had pretty good reason to think that way: science has seemed to push back the frontier of the mysterious and has got us a long way towards mastery of our world, towards greater dignity and freedom. Yet there have always been those who saw that this progress held a seed of arrogance and blindness. It has delegitimized the realm of holiness. It has confused wonder and awe with ignorance.

The new paradigm of science, complex-systems science, doesn't grudgingly leave over that area of the unknown as its own failure, an embarrassment, a barely tolerated shadow. Rather, the self-understanding of science in the new paradigm acknowledges its limitations. Freedom and mystery, wonder and awe, are welcome partners.

---

77 Abraham Joshua Heschel, *Man is Not Alone: A Philosophy of Religion* (New York, Farrar, Straus and Giroux), 1951, 8.

Truth, Humility, and Being Human

The consciousness of "mitzvah" as I'm calling it—the realization that we must do our small part but ultimately don't know how things will turn out—is a core way of thinking that goes back to the beginning of Jewish culture. The ancient rabbis explored questions of the nature of our knowing in their readings of Torah. Often the rabbis would find an inconsistency or a curious or unusual detail in the text of the Torah, a small gap, where something doesn't seem to fit. Instead of explaining these gaps or inconsistencies away as textual errors or the work of different authors badly stitched together by a later redactor, the rabbis would jump right into those gaps and find meanings hidden there. They would play creatively with the intertextual "stringing pearls" that we saw in the Ben Azzai story. So, the story of the creation of humans offered a rich background for this kind of textual exploration.

God says in Genesis 1:26 "Let us make humans in our image…" and people have been asking ever since, "who's the 'us' in this sentence?" We thought the whole point of the Genesis story was to show that One God created the world, not the pantheon of gods that had populated the earlier Mesopotamian creation stories. One of the answers that the rabbis came up with said that God was consulting with the angels about this critical and daring idea of creating a being that would be "in God's image" and also a creature, like other creatures of flesh and blood: a human. Wouldn't this crazy idea bring up all sorts of problems? The angels in this midrash (Beraisheet Rabba 1:8) formed two coalitions: one for and one against:

> Rabbi Simon said, At the time that God came to create the first human the ministering angels formed

themselves into two groups: one said, "Create!" and the other said, "Don't create!" as it is written, (Psalms 85:11) Kindness and Truth met; Justice and Peace kissed...Kindness said, "Create him, because he does much kindness." Truth said, "Don't create him because he is full of lies." Justice said, "Create him, for he acts justly." Peace said, "Don't create him, since he's full of argument." What did God do? God took Truth and threw it to the earth. As it is written, (Daniel 8:12) "And he threw truth to the earth." The angels said to God, Master of the Universe, Why are you insulting your Royal stamp (Truth)? Raise Truth up from the earth! As it is written, "Truth will sprout from the earth." (Psalms 85:12)

Apparently, in order to accomplish the creation of human beings, there would need to be a change in the way truth appears in the world. Humans would need a truth that "will sprout from the earth."

I often pair this midrash with a contemporary story, stringing some pearls to create my own modern midrash:

David Brower was one of the most important founders of the modern environmental movement. His favorite story (as I was told by David Ziv-Krieger, who worked for a while as Brower's assistant back in the 1970s) goes like this: A member of the Cree tribe who was brought into a court to testify in a case involving the damming of a river and the flooding of his ancestral land. This man was asked to swear to tell the Truth, the Whole Truth and nothing but the Truth. He replied that he could not make such an oath to tell the Whole Truth. As a human being all he could possibly do was tell what he knew.

When I juxtapose this story with the midrash about how God created humans, it brings me to wonder about the meaning of

"Truth sprouts from the earth." The truth that this indigenous person recognized included an element of humility and took into account that we humans are parts of a larger, interconnected world that we don't control. Brower's story, I later learned, was not a romantic projection onto indigenous thought, but represented a core way of being in the world that indigenous people often hold. Rupert Ross, who I have cited earlier, also writes of his experiences with Aboriginal (the preferred usage in Canada) people's speech often coming into conflict with the European world view. He describes how Aboriginal speech norms can sometimes be interpreted as disrespectful or lazy in Western ears. He also uses the example of a court of law, where the Aboriginal person might respond to an order by the judge to appear in court on a certain day. They might say "maybe" because it would be considered arrogant to presume that we can know for certain what will happen in the future. Of course, the judges would have heard that as not taking the court order seriously.[78]

This reluctance to be too certain about things reminds me of the old European Jewish custom of saying *"k'ayna hora"* "without the evil eye." This expression is used when mentioning a fortunate circumstance, such as reaching an advanced age, which might, if taken too much for granted, arouse the "evil eye" or misfortune. According to a joke, an elderly Jewish man is called to court as a witness in a case. The judge asks him as a standard preliminary,

"Mr. Bernstein, how old are you?"

Mr. Bernstein answers, "K'ayna hora, 87 years old."

Judge, "Mr. Bernstein, please, just answer the question, "How old are you?"

"K'ayna hora, 87 years old."

"Mr. Bernstein! This could be contempt of court! Please, simply answer the question."

Mr. Bernstein's lawyer now asks to speak, "Judge, if I may, I think I can help here."

---

78 *Returning to the Teachings,* 74.

Judge, "Proceed."

Lawyer, "Mr. Bernstein, k'ayna hora, how old are you?

Mr. Bernstein, "I'm 87."

We laugh at this because it seems to us an old relic of superstition. And perhaps it is. But there is also a taste of that same traditional humility and an awareness of the dangers of arrogance. Those little "superstitions" are reminders of the fact that we can't control everything. Traditional cultures have things like the "evil eye" which punishes that kind of arrogance. Yiddish-speaking European Jews would in my grandparents' time (and some still do) call a beautiful young infant or toddler, "*miskeit*"—which means ugly or disgusting—in order to fool the evil eye. But it is a reminder that many things can happen in life and we cross a dangerous line when we assume to know how things are going to come out.

This can sometimes look like pessimism or fatalism. It's not. There is also a strong current of hope and faith that this longer, more humble view instills.

Hope: Vaçlav Havel and Psalm 37

One of the qualities of "mitzvah"—doing the right thing even if you're not sure how it's all going to come out—is that it fosters an attitude that might be called hope, but is also very close to what is often called faith. The quotation from Vaçlav Havel at the beginning of this chapter well captures that idea: "Hope... is not the same as joy that things are going well, or willingness to invest in enterprises that are obviously headed for early success, but, rather, an ability to work for something because it is good, not just because it stands a chance to succeed."

It is worthwhile to listen to more of his words on this subject:

> The kind of hope I often think about (especially in situations that are particularly hopeless, such as prison) I understand above all as a state of mind, not a state of the world. Either we have hope within

us or we don't; it is a dimension of the soul; it's not essentially dependent on some particular observation of the world or estimate of the situation. Hope is not a prognostication. It is an orientation of the spirit, an orientation of the heart; it transcends the world that is immediately experienced, and is anchored somewhere beyond its horizons.[79]

Havel was able to act under conditions that could easily have smothered hope in any rational estimate of the situation. He was one of a small group of dissidents in the '70s and '80s who dared to stand up to the seemingly all-powerful Communist regime. He was a leader in the Velvet Revolution which finally toppled communism in Czechoslovakia in 1989. He rose to become the first President of the Czech Republic.

This "orientation of the heart," which brings a person to choose to act because of an inner conviction that the action is right, regardless of the outcome, defines the particular mode of being described in this chapter. The idea of mitzvah in Judaism seeks to nurture this same "orientation of the heart." Letting go of the immediate fruits of one's actions and responding to the demand (mitzvah or commandment, or dharma, duty) of the moment brings a spiritual dimension, a dimension of hope or faith, to one's life.

One can see the core of this orientation in biblical literature. Biblical scholar Ellen F. Davis, in her interpretation of Psalm 37 offers a similar call for this orientation of hope under adverse conditions. She situates this psalm as an agrarian protest against the corrupt elites who would rob the villagers of their land and livelihood, and she sees it as a call for hope. She writes, "The psalm seeks to nurture hope in God while calling vividly to mind the elements of a traditional world that is threatened or

---

79 From "An Orientation of the Heart" in Paul Rogat Loeb, *The Impossible Will Take a Little While: A Citizen's Guide to Hope in a Time of Fear* (New York, Basic Books), 2004, 82.

eclipsed, namely, the world of the Israelite village."[80]

One feels the psalmist's words fostering that inner confidence that the righteous, the honest, will eventually be vindicated.

> Do not be vexed by evil men;
>
> Do not be incensed by wrongdoers;
>
> For they soon will wither like grass,
>
> Like verdure fade away.
>
> — Psalm 37:1 (New Jewish Publication Society)

In this psalm there is a powerfully evocative picture of the poor villagers as people of character. It is this quality of individuals and society that makes for sustainability.

> The wicked borrows and does not repay;
>
> The righteous is generous and giving...
>
> He is always generous, and lends,
>
>> and his children are held blessed.
>> — Psalm 37: 21, 26

Davis notes this emphasis on character in the psalm's description of the economic setting: "The poet is clear eyed about the economic situation. 'The wicked borrows and does not repay'—that is the regular practice not just of irresponsible individuals but also of extractive economies. Yet, communities that endure are 'gracious and giving'; they cultivate modest habits of use and accumulation, and with those, the generosity that is often the remarkable grace of the poor:

> Better the little the righteous has
>
>> than the glut of many wicked folk.
>> — Psalm 37:16[81]

---

80 Ellen F. Davis, *Scripture, Culture and Agriculture:* (Cambridge, Cambridge University Press), 2009, 115.
81 Ibid, 116.

Davis brilliantly summarized the psalm in words that remind me of Havel's call for hope and she also invokes the idea of the dynamics of economic, social, and political systems.

> So these are hopeful words for the *anawim*, the "vulnerable," trapped in a killing system that still appears to be strong, though it has already far outreached itself. . . In such a situation, hope cannot mean naïve expectation of personal prosperity, nor even perhaps one's own survival. Rather, it means looking to the inevitable collapse of the system, with the visionary realism that often emerges among the oppressed...[82]

One thinks of Martin Luther King's famous phrase, quoting the 19th century abolitionist minister Theodore Parker, "The arc of the moral universe is long, but it bends toward justice."[83] Yet, while it is true that, as King and Parker describe, it may be a long, gradual bending toward justice, the psalm captures as well one of the qualities of complex systems and change that we are most concerned with here: it can all change in a blink of an eye.

> A little longer and there will be no wicked man;
> You will look at where he was –
> He will be gone.
> But the lowly will inherit the land,
> And delight in abundant well-being.
> — Psalm 37:10

> I saw a wicked man, powerful,
> well-rooted like a robust native tree.

---

82 Ibid, 117.
83 Martin Luther King, Jr., from "Remaining Awake Through a Great Revolution." Speech given at the National Cathedral, March 31, 1968.

Suddenly he vanished and was gone;

I sought him, but he was not to be found.

— Psalm 37:35

As we have seen in so many social and political revolutions from the fall of the U.S.S.R. to the acceptance of same-sex marriage in the U.S., a system can change in a moment: a bifurcation event. It is this quality of complex systems which makes the hope of the oppressed not completely unreasonable. Larger trends can be discerned, but the exact moment of shift can never be predicted. This gives hope that things can—and sometimes do—shift in the blink of an eye.

Why Bother? Climate Change, Gardens, and the Internet

Like Vaçlav Havel in his political struggle with the Communist regime, like the Israelite villagers of Psalm 37 struggling against powerful economic elites, many of us today find ourselves involved in a struggle for the survival of our civilization as we know it. This time the "enemy" is us: our technology and industry which is burning so much fossil fuel that we are changing the climate of the planet.

In April 2008, food and environmental writer Michael Pollan wrote an article in the *New York Times Magazine* entitled, "Why Bother?" He hit upon how a normal citizen, even one concerned about environmental issues, might feel in facing the daunting challenge of combating the global crisis of climate change. One of the things I like about Pollan's work is that he is one of the most eloquent advocates of the kind of complex-system thinking we've been discussing. Pollan describes the climate as a complex system running out of control: ". . . truly terrifying feedback loops threaten to boost the rate of change exponentially, as the shift from white ice to blue water in the Arctic absorbs more sunlight and warming soils everywhere become more biologically active, causing them to release their vast stores of carbon into the air."

His article addresses the perennial problem of the environmental movement: People get discouraged and say, "What can my small actions do against such huge odds? Why bother?" Pollan's answer to "Why bother?" is the flip side of the systems dynamic of the melting polar ice: the chain reaction that can be set off by your small action.

> If you do bother, you will set an example for other people. If enough other people bother, each one influencing yet another in a chain reaction of behavioral change, markets for all manner of green products and alternative technologies will prosper and expand. (Just look at the market for hybrid cars.) Consciousness will be raised, perhaps even changed: new moral imperatives and new taboos might take root in the culture. . . All of this could, theoretically, happen. What I'm describing (imagining would probably be more accurate) is a process of viral social change, and change of this kind, which is nonlinear, is never something anyone can plan or predict or count on.

Pollan even invokes Vaçlav Havel in this article!

> Going personally green is a bet, nothing more or less, though it's one we probably all should make, even if the odds of it paying off aren't great. Sometimes you have to act as if acting will make a difference, even when you can't prove that it will. That, after all, was precisely what happened in Communist Czechoslovakia and Poland, when a handful of individuals like Vaçlav Havel and Adam Michnik resolved that they would simply conduct their lives 'as if' they lived in a free

society. That improbable bet created a tiny space of liberty that, in time, expanded to take in, and then help take down, the whole of the Eastern bloc.

In this article, Pollan makes the seemingly unlikely suggestion that one of the best things people can do to combat climate change is to plant a garden. Gardening saves energy through much less transportation of food, less fossil fertilizer, but even more importantly, it transforms us personally. In planting a garden, you become empowered to see the connections which are lost in our fractured world. You not only start to understand intellectually but also viscerally feel that you're an active part of a complex, interacting network of connections.

In our kitchen, right by the window where it gets a lot of sun, is an avocado sapling, about two feet tall with ten elegant, deep green leaves. It started as an avocado pit which we put into the compost along with all the other kitchen scraps after eating the avocado with our salad or on a piece of toast. I saw the pit growing into a small sapling in the compost heap and rescued it, leaving behind its companion scraps already well on their way to becoming the rich soil of next year's garden. It's a cycle of life and growth that I love to be a part of—partnering with the bacteria, vegetables, soil, sun, and rain to bring some vibrancy and deliciousness to our little corner of the world. It certainly changes me to have this in my life, and I believe that change reverberates out into the wider world.

Pollan's suggestion of taking action to counteract climate change by planting a garden is building on a growing literary tradition: he's drawing on one of his great intellectual and literary forebears and mine, Wendell Berry. Berry's 1979 essay "The Reactor and the Garden" similarly makes the plea for gardening as a much deeper and more complex form of protest than simply picketing a nuclear energy plant (although he does that as well).[84] This is a tradition I'd like to be a part of. Even when our backyard

---

84 In Wendell Berry, *The Gift of Good Land: Further Essays Cultural and Agricultural by Wendell Berry,* (New York, North Point Press), 1981, 161-70.

garden is looking brown and dreary on a late fall day, one of my favorite chores is walking past the garden, out behind the garage to add to our compost piles. Some days it's the first thing that gets me out of the house to feel the cool air, the dew on the grass. Composting is like cooking in reverse. I'm adjusting the mix of warm and cold, wet and dry ingredients, tending the compost to bring about that transformation from already used, smelly leftovers into the rich, dark soil into which I'll plant next year's seeds.

When you do something like gardening that connects you to the natural and social world, you are more likely to feel and see how a small act can have big consequences. For one thing, in a garden, you see with your own eyes how a tiny seed can yield hundreds of tomatoes, dozens of leaves of kale or pods of beans. When you compost you see the magic of transformation from smelly scraps of waste to rich, fragrant soil. Little by little a person can absorb what all those proverbs and traditional wisdom have been saying: do your part, give what you can, connect to the cycle, and the blessings start to flow. Simply starting a garden can be the beginning of replacing a predictable, mechanistic worldview with a living, bountiful, surprising, organic worldview.

This return to the garden, to a more traditional, organically based wisdom shouldn't be confused with an anti-technology bias. We have something today which they didn't have back in Talmudic days or even when Wendell Berry was writing; something which multiplies the viral potential of our small actions in the garden: the Internet. You can do a small action like planting your garden, and you can take a picture on your phone and post it on Facebook, Twitter and Instagram, and you can, with very little effort, make your decisions and your small actions known to thousands of people. You can take that shift in your consciousness that comes from working in your garden and share it with others to build movements supporting regenerative agriculture or supporting activism for changing policy on climate change. The mechanisms for taking the personal action and letting it spread have changed, making it even more plausible that your small mitzvah could be the catalyst for creating real change in the world.

The Mitzvah in the Middle: Embracing Uncertainty While Making Progress

As Michael Pollan, Vaçlav Havel, Wendell Berry, the Psalmist, and others have made clear, it is worth it to "bother." Change can happen; things aren't always what they seem. It can start from below, not just with the elite and powerful. That is one of the characteristics of this shift from linear, mechanical thinking to complex-system thinking: change doesn't necessarily come from the top; it can come from anywhere. "High Culture" isn't any more worthy of study than just plain culture. History happens in the details of what food people buy, how they like to dress, and how they treat their elders and raise their children, as much as in the chronicles of great statesmen, politicians, authors, and generals. Baseball games are won by the player who knows how to hit a single to right when there's a runner on base as much as by the player who hits a lot of homers.[85]

Change can come from anywhere, and a little thing can make a big difference. At the same time, it isn't all about the results. The results may go my way and they may not. Win or lose, it is also a matter of character, of hope or faith or spiritual integrity that I need to do my part. I'm signing this petition because it is the mitzvah, and with the hope and faith that through it I am part of the solution, but not the certainty. I vote for my candidate because it's my civic duty and privilege, my secular mitzvah, as much as I do because I hope my candidate wins.

---

85 As was made famous in the best-selling book and then film *Moneyball* by Michael Lewis in which he describes the use of computers to examine data to find effective players that were otherwise under the radar, thus building up a winning team much more cheaply than would otherwise have been possible. This is another example of how computer technology can sometimes help us to take our attention off the flashy, famous people, and pay attention to the data which tells a more organic story about how every day, unheralded acts can often make a big difference.

To do a mitzvah is to occupy a subtle middle ground between goal-oriented, purposeful action for a cause, and spiritual action, done for the sake of doing the right thing, or the holy thing, but without attachment to the fruits of my action. This middle ground is a kind of spiritual sweet spot. It reflects a larger human condition: we are both actors and acted upon. We are all part of a swirling, pulsing, dance of forces, ideas, and movements. Tectonic movements shift under our feet, literally and figuratively. We cannot control it all. It carries us along and, yet, we can sometimes step into the dance and change its direction. We can hold onto someone's hand and they can grab someone else's hand and we can form a circle, a new group, a movement, a revolution.

Judaism has nurtured a millennia-old yearning for redemption, for a Messiah or at least a Messianic age. The vision of redemption doesn't necessarily look the same for everyone. It has been a vision of a return to our homeland, which has included both the religious dream of rebuilding the ancient Temple in Jerusalem and also the secular Zionist vision of a return to the land, to labor and a socialist utopia. To some Jews redemption has not meant a return to a homeland but rather the creation of an ideal society: it has meant labor unions and political activism, movements for economic, gender, racial equality, peace activism, as well as scientific and medical quests for a better world.

It started with the story of the Exodus. The people who have told and retold that story—both Jews and non-Jews—have been imbued with the idea that we need not be stuck as slaves; we can look to a better world and work towards it. Yet the ancient Talmudic rabbis instruct that if you are planting a tree and you see the Messiah coming, you should finish planting the tree first, then go help usher in the Messiah. It's that sweet middle ground— keep your eye on the big prize, but right now, plant that tree. Do that little, grounded mitzvah—and who knows—maybe you'll change the world.

Conclusion

The story I told earlier in the book about how I met my wife when she called me out of the blue, needing a place for Shabbat, has a prelude which I left out of the first telling. When Ilana decided to make that fateful phone call in May it wasn't completely out of the blue. Six months earlier, back in December, I happened to be in Boston for a conference, staying with a friend, and we came to his favorite synagogue, Temple Beth Zion (usually referred to as TBZ) for a Friday night Shabbat service. I saw her from across the proverbial crowded room, thought she looked familiar, but couldn't quite place where I had met her. The thought popped into my head, "I wonder if she is my *basherte* (Yiddish for soul mate)? After services I approached her. She remembered me from the few times that we had met in the past but she was dating someone else and showed no interest in me. That was that. But perhaps a seed was planted.

The time wasn't right, but as it turned out we needed only a few more months for the stars to align, and when they did, thank God, we grabbed the moment. As human beings, we find ourselves living in a moment when timing is crucial. We are running out of time as the climate crisis clock ticks closer to irreversible disaster. But, as we have written, that is just one aspect of a world that is fragmenting in every sphere: social, spiritual, psychological, political, biological, geological. We look around in fear and dismay as the dynamic patterns that have sustained life on earth for millions of years are breaking down in the Anthropocene—the era of human progress and planetary destruction. We see it in the loss of species, from coral reefs to amphibians to bees; in the increasing destructiveness of storms, hurricanes, droughts, floods and fires. In our societies we see depression, suicide, addiction and obesity, diseases of despair, contributing to the chaos and increasing the rush to personality cults and political saviors.

The *Pearl* and the *Flame*

But, at the same time, we are also seeing a paradigm shift toward people coming together, and seeing the world as an interconnected whole. It's taking place across many fields and areas of life: people starting to re-discover the value of community, starting to shift toward systems approaches, understanding ourselves and our world as filled with relationships, a complex web of patterns. We are starting to make that beautiful, complex movement of the spiral: simultaneously returning to a recovered past, while moving forward into our own new integration. We are awakening to the realization that we find our true life and joy, freedom and creativity as parts of the living patterns, coming together as a manifestation of the One Holy Source of All.

We're seeing the graphs cross before our eyes—X—with one line showing the increasing disintegration and the other our movement toward a healthy re-integration. I'm reminded of the description that I read in Robin Wall Kimmerer's wonderful book *Braiding Sweetgrass*, of the cooperative relationship between fungi and algae to form lichen. These blended beings live where there is no soil, nothing to hold onto except tiny cavities on rocks. The algae take sunlight and air to produce food through photosynthesis while the fungi break down minerals. Together the lichen can endure dry and heat, moist and cold, weathering the harshest of conditions.

When researchers conducted experiments in the lab to try to get some algae and fungus to come together, they found that it wasn't easy. They first tried creating ideal conditions for both fungi and algae, and the two species were happy to live separately. It was only when the researchers created a harsh, difficult environment that the algae and fungi came together into their symbiosis. Kimmerer concludes:

> When times are easy and there's plenty to go around, individual species can go it alone. But when conditions are harsh and life is tenuous, it takes a team sworn to reciprocity to keep life going forward. In a world

198

of scarcity, interconnection and mutual aid become critical for survival. So say the lichens.[86]

As difficult as it is to say it, we may be like that as well. It is only as we feel the results of our dis-connecting, our dissecting of ourselves and our world to the point of collapse, that we start to come together. In a recent article in The Atlantic (January 2020), David Brooks writes about the failed modern experiment of the nuclear family. He draws on the substantial literature which has established that this subset of hyper-individualism, the myth of the nuclear family as an ideal, has failed badly, leaving people lonelier, sicker, poorer and unhappier. Brooks also writes about how, especially since the recession of 2008, there has been a growing trend toward creating new kinds of families. Be it a return to multi-generational households, various forms of co-housing communities or alternative models of the family based on different sexual and gender roles, people have started to come together.

For the past 15 years our family has gone almost every summer to a gathering everyone fondly calls "Dance Camp." It's an annual ten day gathering of several dance communities (with names like "Barefoot Boogie" and "Dance Free") from around New England. Everyone volunteers helping cook the meals, people take dance classes and workshops; they dance far into the night and into the wee hours of the morning. But for us, especially when our kids were younger and needed more supervision, it was a blessed model of how much we all need a village. Our kids could roam around with their friends, often led by a young adult in a "LARP" (Live Action Role Play) adventure with padded swords, costumes and a basic script to guide the action. We knew there were plenty of adults around who knew our kids and so we could get some needed extra sleep or take a swim in the lake. When we experienced this way of living, even for only a week or so out of the year, we would marvel at how anyone ever thought that it was a reasonable idea to raise kids as isolated nuclear

---

86 Kimmerer, *Braiding Sweetgrass*, 272.

families. It just doesn't work. And we're all starting to learn that truth not only in our families but in communities, in nature, in work, in governments. We're starting to realize that we've been on the wrong track and change is not only desired, but it's what we desperately need to make us happier and healthier. And we're realizing that we don't have to reinvent the wheel. Our ancient traditions, our indigenous neighbors, our natural environments, and our own bodies are there as teachers.

Coming back to TBZ: after several years in which I was employed as rabbi of a small, part-time congregation in Connecticut and was therefore away for much of the High Holidays, recently, when that job ended, I was happy to be able to pray with my family and home congregation, TBZ, for all of Rosh HaShana, Yom Kippur and all the way through Simchat Torah. This congregation was founded by a charismatic and progressive rabbi, Moshe Waldoks, who introduced a good dose of spirituality, meditation and humor into what had been an aging and declining Conservative synagogue and made it a home for spiritual seekers. My wife, Ilana was one of those spiritual seekers and had been going there for a couple years before we started dating.

The current senior rabbi, Claudia Kreiman, a spirited woman who grew up in Chile and Argentina, and studied for the rabbinate in Israel, is one of those people who can inspire me to experience God. That is because when she prays, she really prays. She is really feeling it, and expresses her devotion and enthusiasm singing, jumping and arm waving as she leads the congregation in prayer. But, of course, she's not doing it alone. Especially during the High Holidays there are always moments that bring me to tears when I feel the whole room praying. Looking around, seeing the emotion in people's faces, joining my prayer to all of theirs' I can feel a Presence. I can never define exactly what is happening, but I feel that we are bringing God into the room. It's a real minyan— we've created something more than the sum of our parts.

So, I come back to emergence/minyan—that magic that happens when we come together, when there is a flow of energy—it can be in that High Holiday service or it can be when I'm writing, drawing, or teaching a class. These are the nourishing moments that fill our lives with joy and give us meaning. Praying at TBZ during the High Holidays, I reflected on how I've come full circle from my experience at Aish HaTorah so many years ago. There I had an experience of God—vicariously—in that I saw that the men in that Jerusalem yeshivah really believed and I didn't. But their belief was part of a world that I couldn't quite enter or fully embrace. It was a five thousand and something year-old-world where science was not respected, progressive values were scorned and they considered themselves to have the exclusive Truth. To join that world, I would need to split myself in a way that I couldn't do. It was either/or.

But now at TBZ I realized that I had found God in a place where social justice, racial and gender equality and deep ecumenism existed hand in hand with deep spirituality—the presence of God was palpable. I reflected that it has been a long and twisted path, but I've come to a both/and place where my modern and progressive world comes together with my religious and spiritual world. There are still problems a plenty, but it is good to know that my life is a part of a growing integration between true spiritual life, progressive social values, and leading-edge science.

In this book I have offered my contribution to this integration by suggesting some connections, or weaving new midrash, which pulls together core Jewish concepts with cutting edge ecological and systems thinking. The "Three Mems:" Minyan (emergence), Mikdash (nestedness) and Mitzvah (bifurcation events or tipping points), draw from the past to point a way into a better future in which we re-member our place as a part of the holy, diverse Wholeness of creation.

Finally, we arrive at the culmination of the High Holidays, Simchat Torah, which we celebrate at TBZ with deeply meditative circling, singing and holding the Torah led by Rav Claudia's husband, Rabbi Ebn Leader. What are we circling? Not a bonfire. Certainly not a statue or an idol of any kind. We're circling the table where the Torah is read, the *bima*. Essentially it is an empty space. We circle something invisible. But we are dancing holding the Torah scrolls, kissing them as they pass by, spinning and jumping, or walking meditatively, building an energetic circle that emerges from our dance. After we've danced and sung our way through the seven *hakafot*—ritual circlings—we lay the Torah scroll down on the table and read, now filling the center of the circle with words of Torah. We read the very last words at the end of the Five Books, which describes the death of Moses, and then immediately we open another scroll and read from the beginning of the Torah, "When God began creating heaven and earth…"

Going immediately from the ending to the beginning that way expresses the idea that the words of the Torah are also a circle, and they tell a story that we enter into, year in and year out. The words are always the same, but we are different each time we hear them. We take the words into the unique stories of our lives, finding new meanings, new insights, as long as we let the words enter deeply, opening up to them so that, from our lives and the words of Torah a new story emerges. The Torah is the necklace upon which we string together the moments of our lives. Like Ben Azzai nearly two thousand years ago, our intellectual life is also spiritual, woven into ritual on festivals such as Simchat Torah, but also on Passover when we tell the stories of redemption from Egypt over symbolic foods, mixing ritual and mind. And like Ben Azzai, who's love for the Torah is compared to stringing of pearls on the neck of one's beloved, for me the love and joy of Torah arises out of this deep sense of weaving together the world of spirit, body, emotion and community. When this happens, I can feel, even just a little bit, the sparks of those flames that were surrounding Ben Azzai when he strung together the pearls of Torah. And the words of Torah are as happy as when they were given on Mount Sinai, and as joyful as when they were new.

Appendix
Structure and Meaning in Mishnah Megillah, Ch. 1[87]

This is the chapter of the Mishnah which was pivotal for my integration of Jewish learning into my growing awareness of a holistic, systemic way of looking at the world such as I was learning in my anthropology major at Reed College. This chapter of the Mishnah has been understood by academic scholars as having been stitched together by an editor or redactor from two unrelated, previously existing collections of *mishnayot* (plural of mishnah).[88] While I agree that this is probably what happened, the method that I am using goes further and posits that the final editor not only stuck the two units together for mnemonic purposes, but actively edited the pre-existing material into a sophisticated composition in order to explore important cultural and religious questions. The first section has to do with the

---

87 This short essay is largely according to the way that I learned this from Rabbi Dov Berkovits in 1980 at The Pardes Institute in Jerusalem. It was a life-changing class at the time and I am still reaping the benefits of Rabbi Berkovitz's teachings. There have been places where I have come to my own understandings and have been influenced by other rabbis and scholars as well but the (what I consider) brilliant insights into the structure of this chapter are for the most part his. I particularly want to mention Rabbi Avraham Walfish, who was also teaching at Pardes at this time. Walfish was influenced by Berkovits in his method of approaching the Mishnah and took this in his own direction, integrating it with academic scholarship and making major contributions to establishing a literary approach to the Mishnah in the academic study of rabbinic texts. A more complete version of my understanding of this chapter of the Mishnah can be found in my Ph.D. dissertation, *Life Containing Texts: A Literary Anthropological Approach to Gender in the Mishnah*, U.C. Berkeley, 2001.

88 This is a bit confusing because the entire book is called the Mishnah, and each individual unit is also called "a mishnah." Thus this chapter of the work called the Mishnah is composed of eleven *mishnayot* or individual mishnah units.

dates that the Scroll of Esther may be read in different places to celebrate the holiday of Purim. The second and longer section has been seen as a collection based on a mnemonic principle: each of the units contains the phrase "there is no difference between x and y except z." In Hebrew: *eyn bein x ve y eleh z.*

Here is the Mishnah in my own English translation:

Mishnah Megillah, Chapter One

1. The Scroll (of Esther) is read on the eleventh, the twelfth, the thirteenth, the fourteenth or the fifteenth (day of Adar), not earlier and not later. Cities surrounded by a wall since the days of Joshua, son of Nun, read on the fifteenth; villages and large towns read it on the fourteenth. But the villages (sometimes) advance (the day of reading) to the day of assembly.

2. How (did this work)? (When) the fourteenth (of Adar) falls on a Monday, the villages and large towns read on that day and the walled cities read on the following day (the fifteenth). (When) the fourteenth falls on a Tuesday or Wednesday, the villages advance to the day of assembly (Monday); the large towns read on that day (the fourteenth), and the walled cities read on the following day (the fifteenth). (When) the fourteenth falls on Thursday, the villages and large towns read on that day and the walled cities (read) on the following day. (When) the fourteenth falls on Shabbat eve (Friday), the villages advance (their reading) to the day of assembly (Thursday) and the large towns and the walled cities read on that day, (Friday). If the fourteenth falls on Shabbat, both the villages and large towns advance and read on the day of assembly, (Thursday), and the walled cities read on the following day. If the fourteenth falls on Sunday, the villages advance and read on the day of assembly (Thursday); and the large towns read on that day (Sunday); and the walled cities read on the following day.

3. What is considered a large city? Any city in which there are ten idlers (available to make a minyan). If there are less than that, it is a village. In these (the Rabbis) said that one advances (the reading) and one does not postpone. But, in the cases of "priests wood" (priestly family holidays celebrating their turn to donate wood for the altar); and the Ninth of Av; the Festival peace-offering; and the commandment of Assembly (of the Jewish people once in seven years); one postpones (until after Shabbat) and does not advance. Even though they (the Rabbis) said that one advances the time for reading the Megilla and one does not postpone the reading, one is permitted to eulogize and fast and give gifts to the poor (on those days other than the fourteenth or fifteen when one reads); Rabbi Yehuda said: When (under what conditions) does a locale read the Megilla on the day of assembly? In a place where (the villagers) enter (the large towns) on Monday and Thursday. But in a place where they do not enter town on Monday and Thursday, one may read it (the Megilla) only in its (designated) time, (the fourteenth).

4. If they read the Megilla during First Adar and then the year was then intercalated (a month was added by the rabbinic court), one reads the Megilla (again) in Second Adar. There is no difference between First Adar and Second Adar except the reading of the Megilla and distributing gifts to the poor (are done in Second Adar and not in First Adar).

5. There is no difference between Festivals and Shabbat except only preparing food (is allowed on Festivals but not Shabbat). There is no difference between Shabbat and Yom Kippur except that this one's (i.e., Shabbat) intentional desecration is punishable by human hands (i.e., the court punishes the offender) and that one's (i.e., Yom Kippur) intentional desecration is punishable with (God's) cutting off (the person's soul from the people).

6. There is no difference between one for whom benefit from another is forbidden by vow and one for whom benefit from another's food is forbidden by vow except stepping foot on

their property, and borrowing utensils that are not used in the preparation of food, (i.e., the one whose vow prohibits them only from the other's food is allowed these things) There is no difference between (animals dedicated to the Temple as) vow offerings and (animals dedicated as) gift offerings except that in the case of vow offerings (if they died or were lost before being sacrificed) one is obligated in the responsibility (to replace them) and in the case of gift offerings, (if they died or were lost) one is not obligated in the responsibility (to replace them).

7. There is no difference between a *zav* who experiences two emissions of a discharge from his penis and one who experiences three emissions except (that the *zav* who experienced three emissions must bring) an offering. There is no difference between a quarantined leper (one whose symptoms are inconclusive, and so must only quarantine) and a confirmed leper except letting the hair on one's head grow wild and rending one's garments. (The confirmed leper must let their hair grow wild and rend their garments). There is no difference between a leper purified from quarantine and a leper purified from confirmed leprosy except (the confirmed leper is obligated in) shaving their body hair and bringing birds as a purification offering.

8. There is no difference between Torah scrolls, and *tefillin* (phylacteries) and *mezuzot* except that Torah scrolls are written in any language, whereas *tefillin* and *mezuzot* are written only in *Ashurit* (in Hebrew with Hebrew script). Rabban Shimon ben Gamliel says, even for Torah scrolls, (the Rabbis) only permitted them to be written in Greek.

9. There is no difference between a High Priest anointed with the oil of anointing (i.e., during the First Temple period), and one consecrated by donning multiple garments (worn by the High Priest during the Second Temple period), except that only the First Temple High Priest brought the bull that comes for transgression of any of the mitzvot. There is no difference between a High

Priest currently serving and a former High Priest (who is no longer serving), except the bull brought by the High Priest on Yom Kippur, and the tenth of an ephah (the daily meal-offering. Both of these are brought only by the current High Priest).

10. There is no difference between a great altar (i.e., a national, centralized altar) and a small altar (a local altar) except the Passover offering (which is only offered on a great altar). This is the principle: Any offering that is vowed or voluntarily dedicated may be sacrificed on a small altar, and any offering that is neither vowed nor voluntarily dedicated (but rather is compulsory) is not sacrificed on a (small) altar.

11. There is no difference between Shilo and Jerusalem except that in Shiloh one eats offerings of lesser sanctity, e.g., individual peace-offerings, thanks-offerings, and the Passover offering, and also the second tithe, in any place that overlooks Shiloh. And in Jerusalem one eats those items only within the walls. And here (in Shiloh) and there (in Jerusalem) offerings of the most sacred order are eaten only within the hangings. (The Tabernacle courtyard in Shiloh was surrounded by hangings and the Temple courtyard in Jerusalem was surrounded by a wall which was the equivalent to the hangings). Also, the sanctity of Shiloh: after (the Tabernacle in Shilo, which was a national, great altar, was destroyed) there is permission (to sacrifice offerings on local altars). But with regard to the sanctity of Jerusalem, after the Temple was destroyed, there is no permission (to return to sacrificing offerings on local altars).

The "Eyn Ben" Collection

The "*eyn ben*" collection, as it has been called,[89] certainly looks like a bunch of unrelated laws, with nothing much uniting them

---

89 I prefer to call it a series rather than a collection. Collection has been the term used by academics which emphasizes the idea that these are random laws that have been collected here solely because of the term "*eyn ben.*" In this essay I'll use the term series which emphasizes that they form a meaningful unit.

except the use of that phrase. However, our analysis convincingly shows that there was a skilled editorial hand which selected and modified any previously existing collection into a subtle and expressive literary unit. This unit was also skillfully stitched together with the first unit about times for reading the Scroll (*megillah*) of Esther to form a complete literary unit of the chapter.

The issue that is underlying the many varied subjects in our chapter is the transition from biblical modes of worship to a new, rabbinic mode. This was a difficult but necessary paradigm shift that the rabbis needed to make if they were to keep the Jewish religion as they understood it, alive (and also expand and strengthen their own precarious and contested positions as leaders). They needed to change and update in order to keep relevant, but they needed to do so in a way that maintained a sense of continuity with the past.

In this chapter the main technique they use is to go back into the raw material of the earlier, biblical religion and present it in such a way as to highlight examples that pointed to elements of their newer, rabbinic religion. For example, one of the major themes that the rabbis of the Mishnah wished to emphasize was the holiness which is found in people, in communities and individuals. This contrasts with the more prevalent biblical religious approach which emphasized a God-centered holiness, coming down from above, as it were. However, the people-centered holiness was not completely absent from biblical religion and the rabbis deployed some of those examples to make it seem as if they were simply drawing on old, venerated tradition, and not really changing anything.

This theme of human-centered holiness is introduced subtly in the beginning of our chapter with the otherwise curious detail that walled cities from the time of Joshua son of Nun are given a unique status and therefore read the Megillah on the 15th of Adar instead of the 14th. This is strange because in the Scroll of Esther the reader is informed that there was indeed a difference between a capitol (walled) city, Shushan, and the outlying towns and villages in that the Jews vanquished their enemies on the

13th of Adar in the towns and villages and celebrated on the 14th, while in Shushan, they vanquished their enemies on the 14th and celebrated on the 15th. So, why didn't the Mishnah simply follow the text of the Scroll of Esther and say that those walled cities from the time of Shushan celebrate on the 15th?

The Talmud gives an answer indicating that it had to do with honoring the land of Israel, which seems to only indirectly answer the question of why the Mishnah names "from the time of Joshua son of Nun." But we do find something directly mentioning walled cities and Joshua in a midrash (*Sifrei, parshat Naso*) which states that at the time that the Israelites entered the land of Canaan after the exodus and wandering for forty years in the desert, their leader, Joshua son of Nun, transferred the holiness that existed the encampment of the Israelites (*mahane Israel*) onto the walled cities. This reference is reinforced by a Mishnah in tractate Kelim (1:7) which lists ten levels of holiness in the land of Israel, starting with the highest level in the Holy of Holies in the center of the Temple in Jerusalem and including walled cities from the time of Joshua son of Nun.

The holiness of the "encampment of Israel" clearly indicates a holiness that is associated with the people and their gathering together. In the beginning of Mishnah Megillah we see another clear sign of this human centered holiness in the spreading of the reading of the megillah (scroll) not only on the days mentioned in the Scroll of Esther: the 14th and 15th of Adar, but also on the 11th, 12th, and 13th. The reason for allowing this radical change was so that the villagers could read together with the larger towns when they came into the towns on market days.

Biblical holidays had specific events with which they were associated, and specific times which corresponded to those events. Those times were considered not simply opportunities for historical recollection of the events, but a celebration of the spiritual quality of that event which infused that special time year after year. So, the Passover holiday is always the "time of our freedom," the Succot holiday is the "time of our Joy," etc. Shifting the megillah reading, a key element of the celebration of Purim, from "it's time" (*zemanah*) in order to allow the people to gather together in larger groups (in order, in other words, to make a minyan) is a radical concession and powerful statement in favor of the power of human-centered holiness.

The theme of walled cities as a symbol of human-centered holiness reinforces the surprising broadening of the times for reading the megillah and it is important for another reason: walled cities appear again at the very end of the chapter, forming an envelope structure in the whole chapter—and providing strong evidence that the *"eyn ben"* section was purposely tied together with the beginning section.

Going through the entire *"eyn ben"* series is too large a task for this appendix, but I offer here a schematic chart of the basic chiastic structure that can be found to organize this series of mishnayot.

Chiastic Structure of the *Eyn Ben* Series

A. Mishnah 5: *okhel* (food) movement away from Purim (in mishnayot 1- 4), the most human-centered holiday, to biblical holidays, going from the most human-centered holidays to the most God-centered.

B. Mishnah 6: *nedarim, nedavot* (oaths and vows) movement back in time, toward biblical era. Movement toward individual.

C. Mishnah 7 (body and clothing) continues movement back in time, also toward individual, body.

D. Mishnah 8: (books, *tefillin, mizuzot* (texts which are placed on the body or in the home.

C* Mishnah 9: body/clothing (individual), biblical era—now moving foward in time.

B* Mishnah 10: *nedarim, nedavot* (oaths and vows), individual/ commuity /nation movement foward in time

A* Mishnah 11: *okhel* (food), movement back to the present, Rabbinic era, nation (emcampment of Israel as expressed in walled city of Jerusalem).

Explicit verbal cues such as the repetition of *okhel* (food) in the A – A* pair, and *nedarim, nedavot* (oaths and vows) in the B – B* pair as well as the thematic repetition body/clothing in C-C* and the unique wording of Mishnah 8 in the central position of D, all are strong indications of a purposeful chiastic structure.[90]

Building on those explicit literary cues, we can discern broader patterns in this chiastic structure. We are first led back to the biblical paradigm of holidays after the surprise of the first four mishnayot which discussed Purim and a new paradigm of human-centered holiness. However, the biblical holidays here are not randomly listed. The *"eyn ben"* "there is no difference except" structure first sets up a comparison between Yom Tov (biblical pilgrimage holiday such as Passover, Shavuot, and Succot) and Shabbat, and then a comparison between Shabbat and Yom Kippur. The comparisons lead us to notice that there is a progression from most human-centered to most God-centered: Yom Tov—Shabbat—Yom Kippur. The ability to prepare and eat food is used here as a measure of how human-centered a holiday is. On Yom Tov one may even cook to prepare food, on Shabbat one may eat food but not prepare it, and on Yom Kippur one must observe the day fully devoted to God, and one is prohibited from eating at all.

Next, mishnah 6 discusses oaths and vows. These bring us, on the one hand, into the world of God-centered holiness in that one makes an oath or vow on by invoking some holy object related to the Temple by saying, for example "such and such is like a sacrifice to me." Since a sacrifice was wholly dedicated to God it was off limits to anyone else, so this would be a way of saying, "I swear off such and such." On the other hand, it

---

90 This parallel is actually stronger than simply the theme of body/clothing. The reader may recognize from the discussion in Chapter Two about the Priest and the *metzorah* (leper) that these are two polar opposite social positions within the biblical social structure. The Priest is at the center of the social hierarchy, must have unblemished body, may not tear his clothing or let his hair be "wild" (*paruah*), while the leper is on the outside of the society, has blemished skin, torn clothing and wild (*paruah*) hair.

is a subtle reminder that even this biblically ordained practice, which focuses on the Temple and sacrifices, could be initiated and used by any individual. The oaths and vows were in effect saying that an individual person could create a realm of holiness, a realm of restricted access on the model of the Temple, on their own private initiative. This is one example of the ways that the Mishnah prepares the reader for a new, rabbinic era, where individual initiative would become more important, by showing that a precedent existed in biblical times.

Mishnah 7 takes the reader even more deeply back into the biblical era with the laws of the *zav*, one who has a genital emission, and the laws of the *metzorah*, the leper. While the laws of oaths and vows are based in biblical texts and imagery, they were still in force during the rabbinic era. The laws of the *zav* and the *metzorah* were more purely biblical and with few exceptions were no longer in force. As we saw in the previous mishnah this mishnah also focuses on the individual, this time bringing the reader's attention directly to the body and clothing of the individual. The mention of *pri'ah* and *primah*, letting the hair on one's head grow wild and rending one's garments, may remind the reader of the opposite laws which apply to the priests—they must never let their hair grow wild and must never rend their garments. As mentioned, this reinforces a parallelism between C (mishnah 7) and C* (mishnah 9) in the chiastic structure of the *eyn ben* series.

Mishnah 8 is the center of the chiastic structure as well as the turning point of the whole chapter. According to the emerging pattern of moving backward in time to the biblical era, then, on the other side of the chiastic structure, moving in the opposite direction, we might have expected that mishnah 8 would have been the epi-center of that biblical, priestly religion. Perhaps the Holy of Holies in the Temple. However, instead we find books, *tefillin* and *mezuzot*. It is true that two out of the three of these are connected to the body and the individual, which were also themes in the chiastic pattern. *Tefillin* which are leather straps holding boxes containing scriptures which are placed on the arm and head of the wearer. are very much connected to the individual body. *Mezuzot* are similarly boxes containing scriptures. These are placed on the doorposts and gates of the home, again emphasizing, if not the individual body, at least the private realm of the family.

So, what is going on in mishnah 8? In fact, according to the biblical texts, in the Holy of Holies there was a unique Torah text: in the very center of the Holy of Holies was the Ark of the Covenant, which contained the original Tablets of the Ten Commandments. So, the Torah in the biblical age was in a box, serving a purely ritual function in the priestly religion. During the rabbinic era, it now appears that books are to be brought out into the public arena, not only to be ritual objects, but are to be read. The Mishnah goes further and emphasizes that they can be written in any language, implying that not only experts like rabbis would be reading the books of Torah, but that even people who needed a translation were given the opportunity to read and understand the Torah books. The scriptures in the boxes, serving a ritual function, now would be those that are on the bodies of individuals or on the doorposts of their homes.

Thus, mishnah 8 appears to be offering in a skillful literary twist a new "holy of holies" for the rabbinic era: the holy books of Torah which were hidden away for the priests and as ritual objects in biblical religion would now be open and available to the nation as a whole. The ritual scriptures, instead of being in a central Temple, would now be literally on the bodies and in the homes of individual Jews. We now see why the beginning of this chapter made such radical changes to make sure that the Megillah, the Scroll of Esther, would be read in all communities—reading and understanding the books of scripture would now be the glue that holds the people together and keeps them connected to God.

The second half of the chiastic structure, mishnayot 9 through 11, now brings us in the other direction. As mentioned, mishnah 9 deals with the polar opposite of the *metzorah*, the leper: the High Priests. Parallel to mishnah 7, it deals with their bodies and clothing, and if we look closely, takes the reader from the earlier, First Temple period, to the Second Temple period.

Mishnah 10 moves away from the individual and back towards the idea of a national center. It may have been a bit jolting for the readers of the Mishnah to hear about the alternation between one national center and allowing the smaller, regional altars. The dominant memory would have been of the revered Temple in Jerusalem, which had been destroyed a little more than a century before the Mishnah was published. Bringing in the ancient history of biblical periods when smaller, regional altars were allowed would relativize and put the national center of Jerusalem into a new perspective. It once again showed that the biblical period was not monolithic, but included periods—such as they themselves would be experiencing in the rabbinic era—when there was no dominant national center.

Mishnah 11, the last in the chapter, compares the earlier, less dominant national center of Shiloh with Jerusalem, the mourned symbol of national unity and identity. The mention of food is a clue that this is the book-end of the chiastic structure of the *eyn ben* series which started with mishnah 5. But the most interesting thing is the reappearance of the word "walls." This ties the end of the chapter back to the beginning, strongly suggesting that the editor intentionally wove together the *eyn ben* series with the smaller collection of mishnayot about reading the megillah.

Beyond this, the mention of the walls of Jerusalem brings together the themes of the chapter in dramatic ways. In the beginning of the chapter "walled cities from the time of Joshua son of Nun" were introduced as a symbol of the holiness inherent in the people and their gatherings. *Mahane Israel*, the Encampment of Israel was the term for the whole people as they were encamped together in the desert wanderings of the exodus. The walled cities were therefore prime symbols of unity and human-centered holiness. But Jerusalem was also the place of the Great Altar and the Holy of Holies where God's presence was most palpable. It was therefore also a symbol of the God-centered holiness that characterized the earlier paradigm of biblical religion. Therefore, the walls of Jerusalem represent the integration of both human-centered holiness and God-centered holiness. It is here shown as the idealized national center, and as the last line of the chapter says, after Jerusalem was destroyed, there is no going back to the lesser centers. There is only a hope of someday restoring that yearned for integration.

We see in this brief sketch of the first chapter of Mishnah Megillah that reading it with an eye towards its composition reveals a complex cultural and religious statement on the part of the rabbis. They balanced the need for a new paradigm with a feeling of continuity and reverence for the past. Like anthropologists of their own society, they sifted through a varied grab bag of cultural items and wove them into a pattern. This allowed them to fit their new rabbinic religious perspective into the traditional vocabulary of biblical religion and have it feel like a natural evolution. This kind of skillful weaving is not, I believe, a cynical manipulation, but rather the same kind of feeling one's way forward through difficult transitions that we also experience in our shifting cultural moment today. We can only can hope that we possess the same skill in weaving together wisdom from the past with the best ideas of our own days in order to create a flourishing future.

Photo by Thea Breite

Natan Margalit is a rabbi and scholar with 30 years of experience in teaching, writing, organizing and congregational leadership. Raised in Honolulu, as a young adult he spent 12 years in Israel where he received rabbinic ordination. He returned to the U.S. and earned his Ph.D. in Near Eastern Studies at U.C. Berkeley with focus areas in Talmud, Literary Theory, and Anthropology. He has taught at Bard College, the Reconstructionist Rabbinical College, Hebrew College Rabbinical School, and now is chair of the Rabbinic Texts Department at the Aleph Ordination Program. He is also Director of the Earth-Based Judaism track of the A.O.P., and is founder of the non-profit Organic Torah. He lives in Newton, Massachusetts with his wife and two sons.

www.ingramcontent.com/pod-product-compliance
Lightning Source LLC
Chambersburg PA
CBHW031500120626
46545CB00005B/1689